LANGSTROTH ON THE HIVE

EVERY GOOD MOTHER SHOULD BE THE HONORED QUEEN OF A HAPPY FAMILY.

LANGSTROTH ON THE HIVE AND THE HONEY-BEE

THE ORIGINAL BEEKEEPER'S MANUAL

REV. L. L. LANGSTROTH

ANCIENT WISDOM PUBLICATIONS

WOODLAND, CALIFORNIA

Worker.

Drone.

QUEEN.

Langstroth on the Hive
L.L. Langstroth
ISBN-10 1940849020
ISBN-13 978-1-940849-02-7

Contents

Preface

This Treatise on the Hive and the Honey-Bee, is respectfully submitted by the Author, to the candid consideration of those who are interested in the culture of the most useful as well as wonderful Insect, in all the range of Animated Nature. The information which it contains will be found to be greatly in advance of anything which has yet been presented to the English Reader; and, as far as facilities for practical management are concerned, it is believed to be a very material advance over anything which has hitherto been communicated to the Apiarian Public.

Debarred, by the state of his health, from the more appropriate duties of his Office, and compelled to seek an employment which would call him, as much as possible, into the open air, the Author indulges the hope that the result of his studies and observations, in an important branch of Natural History, will be found of service to the Community as well as to himself. The satisfaction which he has taken in his researches, has been such that he has felt exceedingly desirous of interesting others, in a pursuit which, (without any reference to its pecuniary profits,) is capable of exciting the delight and enthusiasm of all intelligent observers. The Creator may be seen in all the works of his hands; but in few more directly than in the wise economy of the Honey-Bee.

"What well appointed commonwealths! where each Adds to the stock of happiness for all; Wisdom's own forums! whose professors teach Eloquent lessons in their vaulted hall! Galleries of art! and schools of industry! Stores of rich fragrance! Orchestras of song! What marvelous seats of hidden alchemy! How oft, when wandering far and erring long, Man might learn truth and virtue from the BEE!" *Bowring.*

The attention of Clergymen is particularly solicited to the study of this branch of Natural History. An intimate acquaintance with the wonders of the Bee-Hive, while it would benefit them in various ways, might lead them to draw their illustrations, more from natural objects and the world around them, and in this way to adapt them better to the comprehension and sympathies of their hearers. It was, we know, the constant practice of our Lord and Master, to illustrate his teachings from the birds of the air, the lilies of the field, and the common walks of life and pursuits of men. Common Sense, Experience and Religion alike dictate that we should follow his example.

L. L. LANGSTROTH.

Greenfield, Mass., May 25, 1853.

So work the Honey Bees,
Creatures, that, by a rule in Nature, teach
The art of order to a peopled kingdom. — *Shakspeare.*

CHAPTER I. Introduction

The present condition of practical bee-keeping in this country, is known to be deplorably low. From the great mass of agriculturists, and others favorably situated for obtaining honey, it receives not the slightest attention. Notwithstanding the large number of patent hives which have been introduced, the ravages of the bee-moth have increased, and success is becoming more and more precarious. Multitudes have abandoned the pursuit in disgust, while many of the most experienced, are fast settling down into the conviction that all the so-called "Improved Hives" are delusions, and that they must return to the simple box or hollow log, and "*take up*" their bees with sulphur, in the old-fashioned way.

In the present state of public opinion, it requires no little courage to venture upon the introduction of a new hive and system of management; but I feel confident that a *new era* in bee-keeping has arrived, and invite the attention of all interested, to the reasons for this belief. A perusal of this Manual, will, I trust, convince them that there is a better way than any with which they have yet been acquainted. They will here find many hitherto mysterious points in the physiology of the honey-bee, clearly explained, and much valuable information never before communicated to the public.

It is now nearly fifteen years since I first turned my attention to the cultivation of bees. The state of my health having compelled me to live more and more in the open air, I have devoted a large portion of my time, of late years, to a careful investigation of their habits, and to a series of minute and thorough experiments in the construction of hives, and the best methods of managing them, so as to secure the largest practical results.

Very early in my Apiarian studies, I procured an imported copy of the work of the celebrated Huber, and constructed a hive on his plan, which furnished me with favorable opportunities of verifying some of his most valuable discoveries; and I soon found that the prejudices existing against him, were entirely unfounded. Believing that his discoveries laid the foundation for a more extended and profitable system of bee-keeping, I began to experiment with hives of various construction.

The result of all these investigations fell far short of my expectations. I became, however, most thoroughly convinced that no hives were fit to be used, unless they furnished *uncommon protection* against *extremes* of *heat* and more especially of COLD. I accordingly discarded all thin hives made of inch stuff, and constructed my hives of *doubled* materials, enclosing a "dead air" space all around.

These hives, although more expensive in the first cost, proved to be much cheaper in the end, than those I had previously used. The bees *wintered* remarkably well in them, and swarmed *early* and with unusual *regularity*. My next step in advance, was, while I secured my surplus honey in the most convenient, beautiful and salable forms, so to facilitate the entrance of the bees into the honey receptacles, as to secure the largest fruits from their labors.

Although I felt confident that my hive possessed some valuable peculiarities, I still found myself unable to remedy many of the casualties to which bee-keeping is liable. I now perceived that no hive could be made to answer my expectations unless it gave me the *complete control of the combs*, so that I might remove any, or all of them at pleasure. The use of the Huber hive had convinced me that with proper precautions, the combs might be removed without *enraging* the bees, and that these insects were capable of being domesticated or *tamed*, to a most surprising degree. A knowledge of these facts was absolutely necessary to the further progress of my invention, for without it, I should have regarded a hive designed to allow of the removal of the combs, as too dangerous in use, to be of any practical value. At first, I used movable slats or bars placed on rabbets in the front and back of the hive. The bees were induced to build their combs upon these bars, and in carrying them down, to fasten them to the sides of the hive. By severing the attachments to the sides, I was able, at any time, to remove the combs suspended from the bars. There was nothing *new* in the use of movable *bars*; the invention being probably, at least, a hundred years old; and I had myself used such hives on Bevan's plan, very early in the commencement of my experiments. The chief peculiarity in my hives, as now constructed, was the facility with which these bars could be removed without enraging the bees, and their combination with my new mode of obtaining the surplus honey.

With hives of this construction I commenced experimenting on a larger scale than ever, and soon arrived at results which proved to be of the very first importance. I found myself able, if I wished it, to *dispense entirely* with *natural swarming,* and yet to multiply colonies with much greater *rapidity* and *certainty* than by the common methods. I could, in a *short time, strengthen my feeble colonies,* and furnish those which had *lost their Queen* with the means of *obtaining another.* If I suspected that any thing was the matter with a hive, I could *ascertain* its *true condition,* by making a thorough examination of every part, and if the *worms had gained a lodgment,* I could quickly *dispossess* them. In short, I could perform all the operations which will be explained in this treatise, and I now believed that bee-keeping could be made *highly profitable,* and as much a matter of *certainty,* as any other branch of rural economy.

I perceived, however, that one thing was *yet* wanting. The *cutting* of the combs from their attachments to the *sides* of the hive, in order to remove them, was attended with much loss of *time* to myself and to the bees, and in order to *facilitate* this operation, the construction of my hive was necessarily *complicated.* This led me to invent a method by which the combs were attached to MOVABLE FRAMES, and suspended in the hives, *so as to touch neither the top, bottom, nor sides.* By this device, I was able to remove the combs at pleasure, and if desired, I could speedily transfer them, bees and all, *without any cutting,* to another hive. I have experimented largely with hives of this construction, and find that they answer most admirably, all the ends proposed in their invention.

While experimenting in the summer of 1851, with some observing hives of a peculiar construction, I discovered that bees could be made to work in glass hives, *exposed to the full light of day.* The notice, in a Philadelphia newspaper, of this discovery, procured me the pleasure of an acquaintance with Rev. Dr. Berg, pastor of a Dutch Reformed church in that city. From him, I first learned that a Prussian clergyman, of the name of Dzierzon, (pronounced Tseertsone,) had attracted the attention of crowned heads, by his important discoveries in the management of bees. Before he communicated the particulars of these discoveries, I explained to Dr. Berg, my system of management, and showed him my hive. He expressed the greatest astonishment at the wonderful similarity in our methods of management, both of us having carried on our investigations without the slightest knowledge of each other's labors. Our hives, he found to differ in some very important respects. In the Dzierzon hive, the combs are not attached to *movable frames,* but to *bars,* so that they cannot, *without cutting,* be removed from the hive. In my hive, which is opened *from the top,* any comb may be taken out, without at all disturbing the others; whereas, in the Dzierzon hive, which is opened from one of the ends, it is often necessary to *cut* and *remove many* combs, in order to get access to a particular one; thus, if the *tenth* comb from the end is to be removed, *nine* combs must be first *cut and taken out*. All this consumes a large amount of time. The German hive does not furnish the surplus honey in a form which would be found most salable in our markets, or which would admit of safe transportation in the comb. Notwithstanding these disadvantages, it has achieved a *great triumph* in Germany, and given a *new impulse* to the cultivation of bees.

The following letter from Samuel Wagner, Esq., cashier of the bank in York, Pennsylvania, will show the results which have been obtained in Germany, by the new system of management, and his estimate of the superior value of my hive to those in use there.

York, Pa., Dec. 24, 1852.

Dear Sir,

The Dzierzon theory and the system of bee-management based thereon, were originally promulgated, *hypothetically*, in the "Eichstadt Bienenzeitung" or Bee-journal, in 1845, and at once arrested my attention. Subsequently, when in 1848, at the instance of the Prussian government, the Rev. Mr. Dzierzon published his "Theory and Practice of Bee Culture," I imported a copy, which reached me in 1849, and which I translated prior to January 1850. Before the translation was completed, I received a visit from my friend, the Rev. Dr. Berg, of Philadelphia, and in the course of conversation on bee-keeping, mentioned to him the Dzierzon theory and system, as one which I regarded as new and very superior, though I had had no opportunity for testing it practically. In February following, when in Philadelphia, I left with him the translation in manuscript—up to which period, I doubt whether any other person in this country had any knowledge of the Dzierzon theory; except to Dr. Berg I had never mentioned it to any one, save in very general terms.

In September, 1851, Dr. Berg again visited York, and stated to me your investigations, discoveries and inventions. From the account Dr. Berg gave me, I felt assured that you had devised substantially the *same system* as that so successfully pursued by Mr. Dzierzon; but how far *your hive* resembled his I was unable to judge from description alone. I inferred, however, several points of difference. The coincidence as to system, and the principles on which it was evidently founded, struck me as exceedingly singular and interesting, because I felt confident that you had no more knowledge of Mr. Dzierzon and his labors, before Dr. Berg mentioned him and his book to you, than Mr. Dzierzon had of you. These circumstances made me very anxious to examine your hives, and induced me to visit your Apiary in the village of West Philadelphia, last August. In the absence of the keeper, as I informed you, I took the liberty to explore the premises thoroughly, opening and inspecting a number of the hives, and noticing the internal arrangement of the parts. The result was, that I came away convinced that though your system was based on the same principles as Dzierzon's, yet that your hive was almost totally different from his, in construction and arrangement; that while the same objects *substantially* are attained by each, your hive is more simple, more convenient, and much better adapted for general introduction and use, since the mode of using it can be more easily taught. Of its ultimate and triumphant success I have no doubt. I sincerely believe that when it comes under the notice of Mr. Dzierzon, he will himself prefer it to his own. It in fact combines all the good properties which a hive ought to possess, while it is free from the complication, clumsiness, *vain whims*, and decidedly objectionable features, which characterize most of the inventions which profess to be at all superior to the simple box, or the common chamber hive.

You may certainly claim *equal credit* with Dzierzon for originality in observation and discovery in the natural history of the honey bee, and for success in deducing principles and devising a most valuable system of management from observed facts. But in *invention*, as far as neatness, compactness, and adaptation of means to ends are concerned, the sturdy German must yield the palm to you. You will find a case of similar coincidence detailed in the Westminster Review for October, 1852, page 267, et seq.

I send you herewith some interesting statements respecting Dzierzon, and the estimate in which his system is held in Germany.

Very truly yours,

SAMUEL WAGNER.

Rev. L. L. Langstroth.
The following are the statements to which Mr. Wagner refers.—
"As the best test of the value of Mr. Dzierzon's system, is the

results which have been made to flow from it, a brief account of its rise and progress maybe found interesting. In 1835 he commenced bee-keeping in the common way, with 12 colonies—and after various mishaps, which taught him the defects of the common hives and the old mode of management, his stock was so reduced that in 1838 he had virtually to begin anew. At this period he contrived his improved hive in its ruder form, which gave him the command over all the combs, and he began to experiment on the theory which observation and study had enabled him to devise. Thenceforward his progress was as rapid as his success was complete and triumphant. Though he met with frequent reverses—about 70 colonies having been stolen from him, sixty destroyed by fire, and 24 by a flood—yet in 1846 his stock had increased to 360 colonies, and he realized from them that year six thousand pounds of honey, besides several hundred weight of wax. At the same time most of the cultivators in his vicinity who pursued the common methods, had fewer hives than they had when he commenced.

In the year 1848, a fatal pestilence, known by the name of "foul brood," prevailed among his bees, and destroyed nearly all his colonies before it could be subdued—only about ten having escaped the malady, which attacked alike the old stocks and his artificial swarms. He estimates his entire loss that year at over 500 *colonies*. Nevertheless he succeeded so well in multiplying by artificial swarms, the few that remained healthy, that in the fall of 1851 his stock consisted of nearly 400 colonies. He must, therefore, have multiplied his stocks more than three fold each year. "

The highly prosperous condition of his colonies is attested by the Report of the Secretary of the Annual Apiarian Convention which met in his vicinity last spring. This Convention, the fourth which has been held, consisted of 112 experienced and enthusiastic bee-keepers from various districts of Germany and neighboring countries, and among them were some who when they assembled were strong opposers of his system.

They visited and personally examined the Apiaries of Mr. Dzierzon. The report speaks in the very highest terms of his success, and of the manifest superiority of his system of management. He exhibited and satisfactorily explained to his visitors his practice and principles; and they remarked, with astonishment, the *singular docility* of his bees, and the thorough control to which they were subjected. After a full detail of the proceedings, the Secretary goes on to say:—

"Now that I have seen Dzierzon's method practically demonstrated, I must admit that it is attended with fewer difficulties than I had supposed. With his hive and system of management it would seem that bees become at once more docile than they are in other cases. I consider his system the simplest and best means of elevating bee-culture to a profitable pursuit, and of spreading it far and wide over the land—especially as it is peculiarly adapted to districts in which the bees do not readily and regularly swarm. His eminent success in re-establishing his stock after suffering so heavily from the devastating pestilence—in short the recuperative power of the system demonstrates conclusively, that it furnishes the best, perhaps the only means of reinstating bee-culture lo a profitable branch of rural economy.

Dzierzon modestly disclaimed the idea of having attained perfection in his hive. He dwelt rather upon the truth and importance of his *theory* and *system* of *management.*"

From the Leipzig Illustrated Almanac—Report on Agriculture for 1846.

"Bee culture is no longer regarded as of any importance in rural economy."

From the same for 1851, and 1853.

"Since Dzierzon's system has been made known an entire revolution in bee culture has been produced. A new era has been created for it, and bee-keepers are turning their attention to it with renewed zeal. The merits of his discoveries are appreciated by the government, and they recommend his system as worthy the attention of the teachers of common schools.

Mr. Dzierzon resides in a poor sandy district of Middle Silesia, which, according to the common notions of Apiarians, is unfavorable to bee-culture. Yet despite of this and of various mishaps, he has succeeded in realizing 900 dollars as the product of his bees in one season!

By his mode of management, his bees yield, even in the poorest years, from 10 to 15 per cent on the capital invested, and where the colonies are produced by the Apiarian's own skill and labor they cost him only about one-fourth the price at which they are usually valued. In ordinary seasons the profit amounts to from 30 to 50 per cent, and in very favorable seasons from 80 to 100 per cent."

In communicating these facts to the public, I have several objects in view. I freely acknowledge that I take an honest pride in establishing my claims as an independent observer; and as having matured by my own discoveries, the same system of bee-culture, as that which has excited so much interest in Germany; I desire also to have the testimony of the translator of Dzierzon to the superior merits of my hive. Mr. Wagner is extensively known as an able German scholar. He has taken all the numbers of the Bee Journal, a monthly periodical which has been published for more than fifteen years in Germany, and is probably more familiar with the state of Apiarian culture abroad, than any man in this country.

I am anxious further to show that the great importance which I attach to my system of management, is amply justified by the success of those who while pursuing the same system with inferior hives, have attained results, which to common bee-keepers, seem almost incredible. Inventors are very prone to form exaggerated estimates of the value of their labors; and the American public has been so often deluded with patent hives, devised by persons ignorant of the most important principles in the natural history of the bee, and which have utterly failed to answer their professed objects, that they are scarcely to be blamed for rejecting every new hive as unworthy of confidence.

There is now a prospect that a Bee Journal will before long, be established in this country. Such a publication has long been needed. Properly conducted, it will have a most powerful influence in disseminating information, awakening enthusiasm, and guarding the public against the miserable impositions to which it has so long been subjected.

Two such journals are now published monthly in Germany, one of which has been in existence for more than 15 years—and their wide circulation has made thousands well acquainted with those principles, which must constitute the foundation of any enlightened and profitable system of culture.

The truth is that while many of the principal facts in the physiology of the honey bee have long been familiar to scientific observers, it has unfortunately happened that some of the most important have been widely discredited. In themselves they are so *wonderful*, and to those who have not witnessed them, often *so incredible*, that it is not at all strange that they have been rejected as fanciful conceits, or bare-faced inventions.

Many persons have not the slightest idea that *every thing* may be *seen* that takes place in a bee-hive. But hives have for many years, been in use, containing only one large comb, enclosed on both sides, by glass. These hives are darkened by shutters, and when opened, the queen is exposed to observation, as well as all the other bees. Within the last two years, I have discovered that with proper precautions, colonies can be made to work in observing hives, without shutters, and exposed continually to the *full light of day*; so that observations may be made at all times, without in the least interrupting the ordinary operations of the bees. By the aid of such hives, some of the most intelligent citizens of Philadelphia have seen in my Apiary, the queen bee depositing her eggs in the cells, and constantly surrounded by an affectionate circle of her devoted children. They have also witnessed, with astonishment and delight, all the steps in the mysterious process of raising queens from eggs which with the ordinary development, would have produced only the common bees. For more than three months, there was not a day in which some of my colonies were not engaged in making new queens to supply the place of those taken from them, and I had the pleasure of exhibiting all the facts to bee-keepers who never before felt willing to credit them. As *all* my hives are so made that each comb can be taken out, and examined at pleasure, those who use them, can obtain from them all the information which they need, and, are no longer forced to take any thing upon trust.

May I be permitted to express the hope that the time is now at hand, when the number of practical observers will be so multiplied, that ignorant and designing men will neither be able to impose their conceits and falsehoods upon the public, nor be sustained in their attempts to depreciate the valuable discoveries of those who have devoted years of observation and experiment to promote the advancement of Apiarian knowledge.

CHAPTER II. THE HONEY BEE CAPA-
BLE OF BEING TAMED OR DOMES-
TICATED TO A MOST SURPRISING
DEGREE.

If the bee had not such a necessary and yet formidable weapon both of offence and defence, multitudes would be induced to enter upon its cultivation, who are now afraid to have any thing to do with it. As the new system of management which I have devised, seems to add to this inherent difficulty, by taking the greatest possible liberties with so irascible an insect, I deem it important to show clearly, in the very outset, how bees may be managed, so that all necessary operations may be performed in an Apiary, without incurring any serious risk of exciting their anger.

Many persons have been unable to control their expressions of wonder and astonishment, on seeing me open hive after hive, in my experimental Apiary, in the vicinity of Philadelphia, removing the combs covered with bees, and shaking them off in front of the hives; exhibiting the queen, transferring the bees to another hive, and, in short, dealing with them as if they were as harmless as so many flies. I have sometimes been asked if the bees with which I was experimenting, had not been subjected to a long course of instruction, to prepare them for public exhibition; when in some cases, the very hives which I was opening, contained swarms which had been brought only the day before, to my establishment.

Before entering upon the natural history of the bee, I shall antici-pate some principles in its management, in order to prepare my readers to receive, without the doubts which would otherwise be very natural, the statements in my book, and to convince them that almost any one favor-ably situated, may safely enjoy the pleasure and profit of a pursuit, which has been most appropriately styled, "the poetry of rural economy;" and that, without being made too familiar with a sharp little weapon, which can most speedily and effectually convert all the poetry into very sorry prose.

The Creator intended the bee for the comfort of man, as truly as he did the horse or the cow. In the early ages of the world, indeed until very recently, honey was almost the only natural sweet; and the promise of "a land flowing with milk and honey," had then a significance, the full force of which it is difficult for us to realize. The honey bee was, therefore, created not merely with the ability to store up its delicious nectar for its own use, but with certain properties which fitted it to be domesticated, and to labor for man, and without which, he would no more have been able to subject it to his control, than to make a useful beast of burden of a lion or a tiger.

One of the peculiarities which constitutes the very foundation, not merely of my system of management, but of the ability of man to domesticate at all so irascible an insect, has never, to my knowledge, been clearly stated as a great and controlling principle. It may be thus expressed.

A honey bee never volunteers an attack, or acts on the offensive, when it is gorged or filled with honey.

The man who first attempted to lodge a swarm of bees in an artificial hive, was doubtless agreeably surprised at the ease with which he was able to accomplish it. For when the bees are intending to swarm, they fill their honey-bags to their utmost capacity. This is wisely ordered, that they may have materials for commencing operations immediately in their new habitation; that they may not starve if several stormy days should follow their emigration; and that when they leave their hives, they may be in a suitable condition to be secured by man.

They issue from their hives in the most peaceable mood that can well be imagined; and unless they are abused, allow themselves to be treated with great familiarity. The hiving of bees by those who understand their nature, could almost always be conducted without the risk of any annoyance, if it were not the case that some improvident or unfortunate ones occasionally come forth without the soothing supply; and not being stored with honey, are filled with the gall of the bitterest hate against all mankind and animal kind in general, and any one who dares to meddle with them in particular. Such radicals are always to be dreaded, for they must vent their spleen on something, even though they lose their life in the act.

Suppose the whole colony, on sallying forth, to possess such a ferocious spirit; no one would ever dare to hive them, unless clad in a coat of mail, at least bee-proof, and not even then, until all the windows of his house were closed, his domestic animals bestowed in some safe place, and sentinels posted at suitable stations, to warn all comers to look out for something almost as much to be dreaded, as a fiery locomotive in full speed. In short, if the propensity to be exceedingly good-natured after a hearty meal, had not been given to the bee, it could never have been domesticated, and our honey would still be procured from the clefts of rocks, or the hollows of trees.

A second peculiarity in the nature of the bee, and one of which I continually avail myself with the greatest success, may be thus stated.

Bees cannot, under any circumstances, resist the temptation to fill themselves with liquid sweets.

It would be quite as easy for an inveterate miser to look with indifference upon a golden shower of double eagles, falling at his feet and soliciting his appropriation. If then we can contrive a way to call their attention to a treat of running sweets, when we wish to perform any operation which might provoke them, we may be sure they will accept it, and under its genial influence, allow us without molestation, to do what we please.

We must always be particularly careful not to handle them roughly, for they will never allow themselves to be pinched or hurt without thrusting out their sting to resent such an indignity. I always keep a small watering-pot or sprinkler, in my Apiary, and whenever I wish to operate upon a hive, as soon as the cover is taken off, and the bees exposed, I sprinkle them gently with water sweetened with sugar. They help themselves with the greatest eagerness, and in a few moments, are in a perfectly manageable state. The truth is, that bees managed on this plan are always glad to see visitors, and you cannot look in upon them too often, for they expect at every call, to receive a sugared treat by way of a peace-offering.

I can superintend a large number of hives, performing every operation that is necessary for pleasure or profit, and yet not run the risks of being stung, which must frequently be incurred in attempting to manage, in the simplest way, the common hives. Those who are timid may, at first, use a bee-dress; though they will soon discard every thing of the kind, unless they are of the number of those to whom the bees have a special aversion. Such unfortunates are sure to be stung whenever they show themselves in the vicinity of a bee-hive, and they will do well to give the bees a very wide berth.

Apiarians have, for many years, employed the smoke of tobacco for subduing their bees. It deprives them, at once, of all disposition to sting, but it ought never to be used for such a purpose. If the construction of the hives will not permit the bees to be sprinkled with sugar water, the smoke of burning paper or rags will answer every purpose, and the bees will not be likely to resent it; whereas when they recover from the effect of the tobacco, they not infrequently remember, and in no very gentle way, the operator who administered the nauseous dose.

Let all your motions about your hives be gentle and slow. Accustom your bees to your presence; never crush or injure them in any operation; acquaint yourself fully with the principles of management detailed in this treatise, and you will find that you have but little more reason to dread the sting of a bee, than the horns of your favorite cow, or the heels of your faithful horse.

CHAPTER III.   THE QUEEN OR MOTHER-BEE, THE DRONES, AND THE WORKERS; WITH VARIOUS HIGHLY IMPORTANT FACTS IN THEIR NATURAL HISTORY.

Bees can flourish only when associated in large numbers, as a colony. In a solitary state, a single bee is almost as helpless as a new-born child; it is unable to endure even the ordinary chill of a cool summer night.

If a strong colony of bees is examined, a short time before it swarms, three different kinds of bees will be found in the hive.

1st. A bee of peculiar shape, commonly called the *Queen Bee*.

2d. Some hundreds, more or less, of large bees called *Drones*.

3d. Many thousands of a smaller kind, called *Workers* or common bees, and similar to those which are seen on the blossoms. A large number of the cells will be found filled with honey and bee-bread; while vast numbers contain eggs, and immature workers and drones. A few cells of unusual size, are devoted to the rearing of young queens, and are ordinarily to be found in a perfect condition, only in the swarming season.

The *Queen-Bee* is the only *perfect female* in the hive, and all the eggs are laid by her. The *Drones* are the *males*, and the *Workers* are *females*, whose ovaries or "egg-bags" are so *imperfectly developed* that they are incapable of breeding, and which retain the instinct of females, only so far as to give the most devoted attention to feeding and rearing the brood.

These facts have all been demonstrated repeatedly, and are as well established as the most common facts in the breeding of our domestic animals. The knowledge of them in their most important bearings, is absolutely essential to all who expect to realize large profits from an improved method of rearing bees. Those who will not acquire the necessary information, if they keep bees at all, should manage them in the old-fashioned way, which requires the smallest amount either of knowledge or skill.

I am perfectly aware how difficult it is to reason with a large class of bee-keepers, some of whom have been so often imposed upon, that they have lost all faith in the truth of any statements which may be made by any one interested in a patent hive, while others stigmatize all knowledge which does not square with their own, as "book-knowledge," and unworthy the attention of practical men.

If any such read this book, let me remind them again, that all my assertions may be put to the test. So long as the interior of a hive, was to common observers, a profound mystery, ignorant and designing men might assert what they pleased, about what passed in its dark recesses; but now, when all that takes place in it, can, *in a few moments*, be exposed to the *full light of day*, and every one who keeps bees, can *see and examine* for himself, the man who attempts to palm upon the community, his own conceits for facts, will speedily earn for himself, the character both of a fool and an impostor.

The Queen Bee, or as she may more properly be called the mother bee, is the common mother of the whole colony. She reigns therefore, most unquestionably, by a divine right, as every mother is, or ought to be, a queen in her own family. Her shape is entirely different from that of the other bees. While she is not near so bulky as a drone, her body is longer, and of a more *tapering*, or sugar-loaf form than that of a worker, so that she has somewhat of a wasp-like appearance. Her wings are much shorter, in proportion, than those of the drone, or worker; the under part of her body is of a golden color, and the upper part darker than that of the other bees. Her motions are usually slow and matronly, although she can, when she pleases, move with astonishing quickness.

No colony can long exist without the presence of this all-important insect. She is just as necessary to its welfare, as the soul is to the body, for a colony without a queen must as certainly perish, as a body without the spirit hasten to inevitable decay.

She is treated by the bees, as every mother ought to be, by her children, with the most unbounded respect and affection. A circle of her loving offspring constantly surround her, testifying, in various ways, their dutiful regard; offering her honey, from time to time, and always, most politely getting out of her way, to give her a clear path when she wishes to move over the combs. If she is taken from them, as soon as they have ascertained their loss, the whole colony is thrown into a state of the most intense agitation; all the labors of the hive are at once abandoned; the bees run wildly over the combs, and frequently, the whole of them rush forth from the hive, and exhibit all the appearance of anxious search for their beloved mother. Not being able anywhere to find her, they return to their desolate home, and by their mournful tones, reveal their deep sense of so deplorable a calamity. Their note, at such times, more especially when they first realize her loss, is of a peculiarly mournful character; it sounds something like *a succession of wails on the minor key*, and can no more be mistaken by the experienced bee-keeper, for their ordinary, happy hum, than the piteous moanings of a sick child can be confounded, by an anxious mother, with its joyous crowings, when overflowing with health and happiness.

I am perfectly aware that all this will sound to many, much more like romance than sober reality; but I have determined, in writing this book, to state facts, however wonderful, just as they are; confident that they will, before long, be universally received, and hoping that the many wonders in the economy of the honey bee will not only excite a wider interest in its culture, but will lead those who observe them, to adore the wisdom of Him who gave them such admirable instincts. I cannot refrain from quoting here, the forcible remarks of an English clergyman, who appears to be a very great enthusiast in bee-culture.

"Every bee-keeper, if he have only a soul to appreciate the works of God, and an intelligence of an inquisitive order, cannot fail to become deeply interested in observing the wonderful instincts, (instincts akin to reason,) of these admirable creatures; at the same time that he will learn many lessons of practical wisdom from their example. Having acquired a knowledge of their habits, not a bee will buzz in his ear, without recalling to him some of these lessons, and helping to make him a wiser and a better man. It is certain that in all my experience, I never yet met with a keeper of bees, who was not a respectable, well-conducted member of society, and a moral, if not a religious man.[1] It is evident, on reflection, that this pursuit, if well attended to, must occupy some considerable share of a man's time and thoughts. He must be often about his bees, which will help to counteract the baneful effect of the village inn. "*Whoever is fond of his bees is fond of his home*," is an axiom of irrefragable truth, and one which ought to kindle in every one's breast, a favorable regard for a pursuit which has the power to produce so happy an influence. The love of home is the companion of many other virtues, which, if not yet developed into actual exercise, are still only dormant, and may be roused into wakeful energy at any moment."

The fertility of the queen bee has been much under-estimated by most writers. It is truly astonishing. During the height of the breeding season, she will often, under favorable circumstances, lay from two to three thousand eggs, a day! In my observing hives, I have seen her lay, at the rate of six eggs a minute! The fecundity of the female of the white ant, is much greater than this, as she will lay as many as sixty eggs a minute! but then her eggs are simply extruded from her body, to be carried by the workers into suitable nurseries, while the queen bee herself deposits her eggs in their appropriate cells.

On the way in which the eggs of the Queen Bee are fecundated.

I come now to a subject of great practical importance, and one which, until quite recently, has been *attended* with apparently insuperable difficulties.

It has been noticed that the queen bee commences laying in the latter part of winter, or early in spring, and long before there are any drones or males in the hive. (See remarks on Drones.) In what way are these eggs impregnated? Huber, by a long course of the most indefatigable observations, threw much light upon this subject. Before stating his discoveries, I must pay my humble tribute of gratitude and admiration, to this wonderful man. It is mortifying to every scientific naturalist, and I might add, to every honest man acquainted with the facts, to hear such a man as Huber abused by the veriest quacks and imposters; while others who have appropriated from his labors, nearly all that is of any value in their works, to use the words of Pope,

"Damn with faint praise, assent with civil leer, And without sneering, teach the rest to sneer."

Huber, in early manhood, lost the use of his eyes. His opponents imagine that in stating this fact, they have thrown merited discredit on all his pretended discoveries. But to make their case still stronger, they delight to assert that he saw every thing through the medium of his servant Francis Burnens, an ignorant peasant. Now this ignorant peasant was a man of strong native intellect, possessing that indefatigable energy and enthusiasm which are so indispensable to make a good observer. He was a noble specimen of a self-made man, and afterwards rose to be the chief magistrate in the village where he resided. Huber has paid the most admirable tribute to his intelligence, fidelity and indomitable patience, energy and skill.

It would be difficult to find, in any language, a better specimen of the true Baconian or *inductive* system of reasoning, than Huber's work upon bees, and it might be studied as a model of the only true way of investigating nature, so as to arrive at reliable results.

Huber was assisted in his investigations, not only by Burnens, but by his own wife, to whom he was engaged before the loss of his sight, and who nobly persisted in marrying him, notwithstanding his misfortune, and the strenuous dissuasions of her friends. They lived for more than the ordinary term of human life, in the enjoyment of uninterrupted domestic happiness, and the amiable naturalist scarcely felt, in her assiduous attentions, the loss of his sight.

Milton is believed by many, to have been a better poet, for his blindness; and it is highly probable that Huber was a better Apiarian, for the same cause. His active and yet reflective mind demanded constant employment; and he found in the study of the habits of the honey bee, full scope for all his powers. All the facts observed, and experiments tried by his faithful assistants, were daily reported to him, and many inquiries were stated and suggestions made by him, which would probably have escaped his notice, if he had possessed the use of his eyes.

Few have such a command of both time and money as to enable them to carry on, for a series of years, on a grand scale, the most costly experiments. Apiarians owe more to Huber than to any other person. I have repeatedly verified the most important of his observations, and I take *the greatest delight* in acknowledging my obligations to him, and in holding him up to my countrymen, as the Prince of Apiarians.

My Readers will pardon this digression. It would have been morally impossible for me to write a work on bees, without saying at least as much as this, in vindication of Huber.

I return to his discoveries on the impregnation of the Queen Bee. By a long course of experiments most carefully conducted, he ascertained that like many other insects, she is fecundated in the open air, and on the wing, and that the influence of this lasts for several years, and probably for life. He could not form any satisfactory conjecture, as to the way in which the eggs which were not yet developed in her ovaries, could be fertilized. Years ago, the celebrated Dr. John Hunter, and others, supposed that there must be a permanent receptacle for the male sperm, opening into the passage for the eggs called the oviduct. Dzierzon, who must be regarded as one of the ablest contributors of modern times, to Apiarian science, maintains this opinion, and states that he has found such a receptacle filled with a fluid, resembling the semen of the drones. He nowhere, to my knowledge, states that he ever made microscopic examinations, so as to put the matter on the footing of demonstration.

In January and February of 1852, I submitted several Queen Bees to Dr. Joseph Leidy of Philadelphia, for a scientific examination. I need hardly say to any Naturalist in this country, that Dr. Leidy has obtained the very highest reputation, both at home and abroad, as a skillful naturalist and microscopic anatomist. No man in this country or Europe, was more competent to make the investigations that I desired. He found in making his dissections, a small globular sac, not larger than a grain of mustard seed, (about 1/33 of an inch in diameter,) communicating with the oviduct, and filled with a whitish fluid, which when examined under the microscope, was found to abound in spermatozoa, or the animalculæ, which are the unmistakable characteristics of the seminal fluid. Later in the season, the same substance was compared with some taken from the drones, and found to be exactly similar to it.

These examinations have settled, on the impregnable basis of demonstration, the mode in which the eggs of the Queen are vivified. In descending the oviduct to be deposited in the cells, they pass by the mouth of this seminal sac or spermatheca, and receive a portion of its fertilizing contents. Small as it is, its contents are sufficient to impregnate hundreds of thousands of eggs. In precisely the same way, the mother wasps and hornets are fecundated. The females alone of these insects survive the winter, and they begin, single-handed, the construction of a nest, in which, at first, only a few eggs are deposited. How could these eggs hatch, if the females which laid them, had not been impregnated, the previous season? Dissection proves them to have a spermatheca, similar to that of the Queen Bee.

Of all who have written against Huber, no one has treated him with more unfairness, misrepresentation, and I might almost add, malignity, than Huish. He maintains that the eggs of the Queen are impregnated by the drones, after she has deposited them in the cells, and accounts for the fact that brood is produced in the Spring, long before the existence of any drones in the hive, by asserting that these eggs were deposited and impregnated late in the previous season, and have remained dormant, all winter, in the hive: and yet the same writer, while ridiculing the discoveries of Huber, advises that all the mother wasps should be killed in the Spring, to prevent them from founding families to commit depredations upon the bees! It never seems to have occurred to him, that the existence of a permanently impregnated mother wasp, was just as difficult to be accounted for, as the existence of a similarly impregnated Queen Bee.

EFFECT OF RETARDED IMPREGNATION ON THE QUEEN BEE.

I shall now mention a fact in the physiology of the Queen Bee, more singular than any which has yet been related.

Huber, while experimenting to ascertain how the Queen was fecundated, confined some of his young Queens to their hives, by contracting the entrances, so that they were not able to go in search of the drones, until three weeks after their birth. To his amazement, these Queens whose impregnation was thus unnaturally retarded, *never laid any eggs but such as produced drones*!!

He tried the experiment again and again, but always with the same result. Some Bee-Keepers, long before his time, had observed that all the brood in a hive were occasionally drones, and of course, that such colonies rapidly went to ruin. Before attempting any explanation of this astonishing fact, I must call the attention of the reader, to another of the mysteries of the Bee-Hive,

FERTILE WORKERS.

It has already been remarked, that the workers are proved by dissection to be females, all of which, under ordinary circumstances, are barren. Occasionally, some of them appear to be more fully developed than common, so as to be capable of laying eggs: these eggs, like those of Queens whose impregnation has been retarded, *always produce drones*! Sometimes, when a colony has lost its Queen, these drone-laying workers are exalted to her place, and treated with equal respect and affection, by the bees. Huber ascertained that these fertile workers were generally reared in the neighborhood of the young Queens, and he thought that they received some particles of the peculiar food or jelly on which the Queens are reared. (See Royal Jelly.) He did not pretend to account for the effect of retarded impregnation; and made no experiments to determine the facts, as to the fecundation of these fertile workers.

Since the publication of Huber's work, nearly 50 years ago, no light has been shed upon the mysteries of drone-laying Queens and workers, until quite recently. Dzierzon appears to have been the first to ascertain the truth on this subject; and his discovery must certainly be ranked as unfolding one of the most astonishing facts in all the range of animated nature. This fact seems, at first view, so absolutely incredible, that I should not dare to mention it, if it were not supported by the most indubitable evidence, and if I had not, (as I have already observed,) determined to state all important and well ascertained facts, without seeking, by any concealments, to pander to the prejudices of conceited, and often, very ignorant Bee-Keepers.

Dzierzon advances the opinion that impregnation is not needed in order that the eggs of the Queen may produce drones; but, that all impregnated eggs produce females, either workers or Queens; and all unimpregnated ones, males or drones! He states that he found drone-laying Queens in several of his hives, whose wings were so imperfect that they could not fly, and that on examination, they proved to be unfecundated. Hence he concluded that the eggs of the Queen Bee or fertile worker, had from the previous impregnation of the egg which produced them, sufficient vitality to produce the drone, which is a less highly organized insect, and one inferior to the Queen or workers. It had long been known, that the Queen deposits drone eggs in the large or drone cells, and worker eggs in the small or worker cells, and that she makes no mistakes. Dzierzon inferred, therefore, that there was some way in which she was able to decide as to the sex of the egg before it was laid, and that she must have a control over the mouth of the seminal sac, so as to be able to extrude her eggs, allowing them to receive or not, just as she pleased, a portion of its fertilizing contents. In this way he thought she determined the sex, according to the size of the cells in which she laid them. Mr. Samuel Wagner of York, Pa., has recently communicated to me a very original and exceedingly ingenious theory of his own, which he thinks will account for all the facts without admitting that the Queen Bee has any special knowledge or will on the subject. He supposes that when she deposits her eggs in the worker cells, her body is slightly compressed by the size of the cells, and that the eggs, as they pass the spermatheca, receive in this manner, its vivifying influence. On the contrary, when she is egg-laying in drone cells, this compression cannot take place, the mouth of the spermatheca is kept closed, and the eggs are, necessarily, unfecundated. This theory may prove to be true, but at present, it is encumbered with some difficulties and requires further investigation, before it can be considered as fully established.

Leaving then the question whether the Queen exercises any volition in this matter, for the present undecided, I shall state some facts which occurred in the summer of 1852, in my own Apiary, and shall then endeavor to relieve, as far as possible, this intricate subject from some of the difficulties which embarrass it.

In the Autumn of 1852, my assistant found, in one of my hives, a young Queen, the whole of whose progeny was drones. The colony had been formed by removing part of the combs containing bees, brood and eggs from another hive. It had only a few combs, and but a small number of bees. They raised a new Queen in the manner which will hereafter be particularly described. This Queen had laid a number of eggs in one of the combs, and the young bees from some of them were already emerging from the cells. I perceived, at the first glance, that they were drones. As there were none but worker cells in the hive, they were reared in them, and not having space for full development, they were dwarfed in size, although the bees, in order to give them more room, had pieced out the cells so as to make them larger than usual! Size excepted, they appeared as perfect as any other drones.

I was not only struck with the singularity of finding drones reared in worker cells, but with the equally singular fact that a young Queen, who at first lays only the eggs of workers, should be laying drone eggs at all; and at once conjectured that this was a case of a drone-laying, unimpregnated Queen, as sufficient time had not elapsed for her impregnation to be unnaturally retarded. I saw the great importance of taking all necessary precautions to determine this point. The Queen was removed from the hive, and carefully examined. Her wings, although they appeared to be perfect, were so paralized that she could not fly. It seemed probable, therefore, that she had never been able to leave the hive for impregnation.

To settle the question beyond the possibility of doubt, I submitted this Queen to Dr. Joseph Leidy for microscopic examination. The following is an extract from his report: "The ovaries were filled with eggs; the poison sac was full of fluid, and I took the whole of it into my mouth; the poison produced a strong metallic taste, lasting for a considerable time, and at first, it was pungent to the tip of the tongue. The spermatheca was distended with a perfectly colorless, transparent, viscid liquid, *without a trace of spermatozoa."*

This examination seems perfectly to sustain the theory of Dzierzon, and to demonstrate that Queens do not need to be impregnated, in order to lay the eggs of males.

I must confess that very considerable doubts rested on my mind, as to the accuracy of Dzierzon's statements on this subject, and chiefly because of his having hazarded the unfortunate conjecture that the place of the poison bag in the worker, is occupied in the Queen, by the spermatheca. Now this is so completely contrary to fact, that it was a very natural inference that this acute and thoroughly honest observer, made no microscopic dissections of the insects which he examined. I consider myself peculiarly fortunate in having enjoyed the benefit of the labors of a Naturalist, so celebrated as Dr. Leidy, for microscopic dissections. The exceeding minuteness of some of the insects which he has completely figured and described, almost passes belief.

On examining this same colony a few days later, I obtained the most satisfactory evidence that these drone eggs were laid by the Queen which had been removed. No fresh eggs had been deposited in the cells, and the bees, on missing her, had commenced the construction of royal cells, to rear if possible, another Queen, a thing which they would not have done, if a fertile worker had been present, by which the drone eggs had been laid.

Another very interesting fact proves that *all* the eggs laid by this Queen, were drone eggs. Two of the royal cells were, in a short time, discontinued, and were found to be empty, while a third contained a worm, which was sealed over the usual way, to undergo its changes from a worm to a perfect Queen.

I was completely at a loss to account for this, as the bees having an unimpregnated drone-laying Queen, ought not to have had a single female egg from which they could rear a Queen.

At first I imagined that they might have *stolen* it from another hive, but when I opened this cell, it contained, instead of a queen, *a dead drone!*

I then remembered that Huber has described the same mistake on the part of some of his bees. At the base of this cell, was an extraordinary quantity of the peculiar jelly or paste, which is fed to the young that are to be transformed into queens. The poor bees in their desperation, appear to have dosed the unfortunate drone to death: as though they expected by such liberal feeding, to produce some hopeful change in his sexual organization!

It appears to me that these facts constitute all the links in a perfect chain, and demonstrate beyond the possibility of doubt, that unfecundated queens are not only capable of laying eggs, (this would be no more remarkable than the same occurrence in a hen,) but that these eggs are possessed of sufficient vitality to produce drones. Aristotle, who flourished before the Christian era, had noticed that there was no difference in appearance, between the eggs producing drones and those producing workers; and he states that drones only are produced in hives which have no queen; of course the eggs producing them, were laid by fertile workers. Having now the aid of powerful microscopes, we are still unable to detect the slightest difference in size or appearance in the eggs, and this is precisely what we should expect if the same egg will produce either a worker or a drone, according as it is or is not impregnated. The theory which I propose, will, I think, perfectly harmonize with all the observed facts on this subject.

I believe that after fecundation has been delayed for about three weeks, the mouth of the spermatheca becomes permanently closed, so that impregnation can no longer be effected; just as the parts of a flower, after a certain time, wither and shut up, and the plant is incapable of fructification. The fertile drone-laying workers, are in my opinion, physically incapable of being impregnated. However strange it may appear, or even improbable, that an unimpregnated egg can give birth to a living being, or that the sex can be dependent on impregnation, we are not at liberty to reject facts, because we cannot comprehend the reasons of them. He who allows himself to be guilty of such folly, if he seeks to maintain his consistency, will be plunged, sooner or later, into the dreary gulf of atheism. Common sense, philosophy and religion alike teach us to receive all undoubted facts in the natural and the spiritual world, with becoming reverence; assured that however mysterious to us, they are all most beautifully harmonious and consistent in the sight of Him whose "understanding is infinite."

There is something analogous to these wonders in the bee, in what takes place in the aphides or green lice which infest our rose bushes and other plants. We have the most undoubted evidence that a fecundated female gives birth to other females, and they in turn to others still, all of which, without impregnation, are able to bring forth young, until at length, after a number of generations, perfect males and females are produced, and the series starts anew!

The unequaled facilities, furnished by my hives, have seemed to render it peculiarly incumbent on me, to do all in my power to clear up the difficulties in this intricate and yet highly important branch of Apiarian knowledge. All the leading facts in the breeding of bees ought to be as well known to the bee keeper, as the same class of facts in the rearing of his domestic animals. A few crude and hasty notions, but half understood and half digested, will answer only for the old fashioned bee keeper, who deals in the brimstone matches. He who expects to conduct bee keeping on a safe and profitable system, must learn that on this, as on all other subjects, "knowledge is power."

The extraordinary fertility of the queen bee has already been noticed. The process of laying has been well described by the Rev. W. Dunbar, a Scotch Apiarian.

"When the queen is about to lay, she puts her head into a cell, and remains in that position for a second or two, to ascertain its fitness for the deposit which she is about to make. She then withdraws her head, and curving her body downwards,[2] inserts the lower part of it into the cell: in a few seconds she turns half round upon herself and withdraws, leaving an egg behind her. When she lays a considerable number, she does it equally on each side of the comb, those on the one side being as exactly opposite to those on the other as the relative position of the cells will admit. The effect of this is to produce the utmost possible concentration and economy of heat for developing the various changes of the brood!"

Here as at every step in the economy of the bee our minds are filled with admiration as we witness the perfect adaptation of means to ends. Who can blame the warmest enthusiasm of the Apiarian in view of a sagacity which seems scarcely inferior to that of man.

"The eggs of bees," I quote from the admirable treatise of Bevan, "are of a lengthened oval shape, with a slight curvature, and of a bluish white color: being besmeared at the time of laying, with a glutinous substance,[3] they adhere to the bases of the cells, and remain unchanged in figure or situation for three or four days; they are then hatched, the bottom of each cell presenting to view a small white worm. On its growing so as to touch the opposite angle of the cell, it coils itself up, to use the language of Swammerdam, like a dog when going to sleep; and floats in a whitish transparent fluid, which is deposited in the cells by the nursing-bees, and by which it is probably nourished; it becomes gradually enlarged in its dimensions, till the two extremities touch one another and form a ring. In this state it is called a larva or worm. So nicely do the bees calculate the quantity of food which will be required, that none remains in the cell when it is transformed to a nymph. It is the opinion of many eminent naturalists that farina does not constitute the sole food of the larva, but that it consists of a mixture of farina, honey and water, partly digested in the stomachs of the nursing-bees."

"The larva having derived its support, in the manner above described, for four, five or six days, according to the season," (the development being retarded in cool weather, and badly protected hives,) "continues to increase during that period, till it occupies the whole breadth and nearly the length of the cell. The nursing bees now seal over the cell, with a light *brown cover*, externally more or less *convex*, (the cap of a drone cell is more convex than that of a worker,) and thus differing from that of a honey cell which is *paler* and somewhat *concave*." The cap of the brood cell appears to be made of a mixture of bee-bread and wax; it is not air tight as it would be if made of wax alone; but when examined with a microscope it appears to be reticulated, or full of fine holes through which the enclosed insect can have air for all necessary purposes. From its texture and shape it is easily thrust off by the bee when mature, whereas, if it consisted wholly of wax, the young bee would either perish for lack of air, or be unable to force its way into the world! Both the material and shape of the lids which seal up the honey cells are different, because an entirely different object was aimed at; they are of pure wax to make them air tight and thus to prevent the honey from souring or candying in the cells! They are concave or hollowed inwards to give them greater strength to resist the pressure of their contents!

To return to Bevan. "The larva is no sooner perfectly inclosed than it begins to line the cell by spinning round itself, after the manner of the silk worm, a whitish silky film or cocoon, by which it is encased, as it were, in a pod. When it has undergone this change, it has usually borne the name of *nymph* or *pupa*. The insect has now attained its full growth, and the large amount of nutriment which it has taken serves as a store for developing the perfect insect."

"The *working bee nymph* spins its cocoon in thirty-six hours. After passing about three days in this state of preparation for a new existence, it gradually undergoes so great a change as not to wear a vestige of its previous form, but becomes armed with a firmer mail, and with scales of a dark brown hue. On its belly six rings become distinguishable, which by slipping one over another enables the bee to shorten its body whenever it has occasion to do so.

"When it has reached the twenty-first day of its existence, counting from the moment the egg is laid, it comes forth a perfect winged insect. The cocoon is left behind, and forms a closely attached and exact lining to the cell in which it was spun; by this means the breeding cells become smaller and their partitions stronger, the oftener they change their tenants; and may become so much diminished in size as not to admit of the perfect development of full sized bees."

"Such are the respective stages of the working bee: those of the royal bee are as follows: she passes three days in the egg and is five a worm; the workers then close her cell, and she immediately begins spinning her cocoon, which occupies her twenty four hours. On the tenth and eleventh days and a part of the twelfth, as if exhausted by her labor, she remains in complete repose. Then she passes four days and a part of the fifth as a nymph. It is on the sixteenth day therefore that the perfect state of queen is attained."

"The drone passes three days in the egg, six and a half as a worm, and changes into a perfect insect on the twenty-fourth or twenty-fifth day after the egg is laid."

"The *development* of *each species* likewise proceeds more slowly when the colonies are weak or the air cool, and when the weather is very cold it is entirely suspended. Dr. Hunter has observed that the eggs, worms and nymphs all require a heat above 70° of Fahrenheit for their evolution."

In the chapter on protection against extremes of *heat* and *cold*, I have dwelt, at some length, upon the importance of constructing the hives in such a manner as to enable the bees to preserve, as far as possible, a uniform temperature in their tenement. In thin hives exposed to the sun, the heat is sometimes so great as to destroy the eggs and the larvæ, even when the combs escape from being melted; and the cold is often so severe as to check the development of the brood, and sometimes to kill it outright.

In such hives, when the temperature out of doors falls suddenly and severely, the bees at once feel the unfavorable change; they are obliged in self defence to huddle together to keep warm, and thus large portions of the brood comb are often abandoned, and the brood either destroyed at once by the cold, or so enfeebled that they never recover from the shock. Let every bee keeper, in all his operations, remember that brood comb must never be exposed to a low temperature so as to become chilled: the disastrous effects are almost as certain, as when the eggs of a setting hen are left, for too long a time, by the careless mother. The brood combs are never safe when taken for any considerable time from the bees, unless the temperature is fully up to summer heat.

"[4]The young bees break their envelope with their teeth, and assisted, as soon as they come forth, by the older ones, proceed to cleanse themselves from the moisture and exuviæ with which they were surrounded. Both drones and workers on emerging from the cell are, at first grey, soft and comparatively helpless so that some time elapses before they take wing.

"With respect to the cocoons spun by the different larvæ, both workers and drones spin *complete cocoons*, or inclose themselves on every side; royal larvæ construct only *imperfect cocoons*, open behind, and enveloping only the head, thorax, and first ring of the abdomen; and Huber concludes, without any hesitation, that the final cause of their forming only incomplete cocoons is, that they may thus be exposed to the mortal sting of the first hatched queen, whose instinct leads her instantly to seek the destruction of those who would soon become her rivals.

"If the royal larvæ spun complete cocoons, the stings of the queens seeking to destroy their rivals might be so entangled in their meshes that they could not be disengaged. 'Such,' says Huber, 'is the instinctive enmity of young queens to each other, that I have seen one of them, immediately on its emergence from the cell, rush to those of its sisters, and tear to pieces even the imperfect larvæ. Hitherto philosophers have claimed our admiration of nature for her care in preserving and multiplying the species. But from these facts we must now admire her precautions in exposing certain individuals to a mortal hazard.'"

The cocoon of the royal larva is very much stronger and coarser than that spun by the drone or worker, its texture considerably resembling that of the silk worm's. The young queen does not come forth from her cell until she is quite mature; and as its great size gives her abundant room to exercise her wings she is capable of flying as soon as she quits it. While still in her cell she makes the fluttering and piping noises with which every observant bee keeper is so well acquainted.

Some Apiarians have supposed that the queen bee has the power to regulate the development of eggs in her ovaries, so that few or many are produced, according to the necessities of the colony. This is evidently a mistake. Her eggs, like those of the domestic hen, are formed without any volition of her own, and when fully developed, must be extruded. If the weather is unfavorable, or if the colony is too feeble to maintain sufficient heat, a smaller number of eggs are developed in her ovaries, just as unfavorable circumstances diminish the number of eggs laid by the hen; if the weather is very cold, egg-laying usually ceases altogether. In the latitude of Philadelphia, I opened one of my hives on the 5th day of February, and found an abundance of eggs and brood, although the winter had been an unusually cold one, and the temperature of the preceding month very low. The fall of 1852 was a warm one, and eggs and brood were found in a hive which I examined on the 21st of October. Powerful stocks in well protected hives contain some brood, at least ten months in the year; in warm countries, bees probably breed, every month in the year.

It is highly interesting to see in what way the supernumerary eggs of the queen are disposed of. When the number of workers is too small to take charge of all her eggs, or when there is a deficiency of bee bread to nourish the young, (See chapter on Pollen,) or when, for any reason, she judges it not best to deposit them in cells, she stands upon a comb, and simply extrudes them from her oviduct, and the workers devour them as fast as they are laid! This I have repeatedly witnessed in my observing hives, and admired the sagacity of the queen in economizing her necessary work after this fashion, instead of laboriously depositing the eggs in cells where they are not wanted. What a difference between her wise management and the stupidity of a hen obstinately persisting to set upon addled eggs, or pieces of chalk, and often upon nothing at all.

The workers eat up also all the eggs which are dropped, or deposited out of place by the queen; in this way, nothing goes to waste, and even a tiny egg is turned to some account. Was there ever a better comment upon the maxim? "Take care of the pence, and the pounds will take care of themselves."

Do the workers who appear to be so fond of a tit-bit in the shape of a new laid egg, ever experience a struggle between their appetites and the claims of duty, and does it cost them some self denial to refrain from making a breakfast on a fresh laid egg? It is really very difficult for one who has carefully watched the habits of bees, to speak of his little favorites in any other way than as though they possessed an intelligence almost, if not quite, akin to reason.

It is well known to every breeder of poultry, that the fertility of a hen decreases with age, until at length, she becomes entirely barren; it is equally certain that the fertility of the queen bee ordinarily diminishes after she has entered upon her third year. She sometimes ceases to lay Worker eggs, a considerable time before she dies of old age; the contents of the spermatheca are exhausted; the eggs can no longer be impregnated and must therefore produce drones.

The queen bee usually dies of old age, some time in her fourth year, although instances are on record of some having survived a year longer. It is highly important to the bee keeper who would receive the largest returns from his bees, to be able, as in my hives, to catch the queen and remove her, when she has passed the period of her greatest fertility. In the sequel, full directions will be given, as to the proper time and mode of effecting it.

Before proceeding farther in the natural history of the queen bee, I shall describe more particularly, the other inmates of the hive.

THE DRONES OR MALE BEES.

The drones are, unquestionably, the male bees. Dissection proves that they have the appropriate organs of generation. They are much larger and stouter than either the queen or workers; although their bodies are not quite so long as that of the queen. They have no sting with which to defend themselves; no proboscis which is suitable for gathering honey from the flowers, and no baskets on their thighs for holding the bee-bread. They are thus physically disqualified for work, even if they were ever so well disposed to it. Their proper office is to impregnate the young queens, and they are usually destroyed by the bees, soon after this is completed.

Dr. Evans the author of a beautiful poem on bees thus appropriately describes them:—

"Their short proboscis sips No luscious nectar from the wild thyme's lips, From the lime's leaf no amber drops they steal, Nor bear their grooveless thighs the foodful meal: On other's toils in pamper'd leisure thrive The lazy fathers of the industrious hive."

The drones begin to make their appearance in April or May; earlier or later, according to climate and the forwardness of the season, and strength of the stock. They require about twenty-four days for their full development from the egg. In colonies which are too weak to swarm, none, as a general rule, are reared: they are not needed, for in such hives, as no young queens are raised, they would be only useless consumers.

The number of drones in a hive is often very great, amounting, not merely to hundreds, but sometimes to thousands. It seems, at first, very difficult to understand why there should be so many, especially since it has been ascertained that a single one will impregnate a queen for life. But as intercourse always takes place high in the air, the young queens are obliged to leave the hive for this purpose; and it is exceedingly important to their safety, that they should be sure of finding one, without being compelled to make frequent excursions. Being larger than a worker, and less quick on the wing, they are more exposed to be caught by birds, or blown down and destroyed by sudden gusts of wind.

In a large Apiary, a few drones in each hive, or the number usually found in one, might be amply sufficient. But it must be borne in mind, that under these circumstances, bees are not in a state of nature. Before they were domesticated, a colony living in a forest, often had no neighbors for miles. Now a good stock in our climate, sometimes sends out three or more swarms, and in the tropical climates, of which the bee is a native, they increase with astonishing rapidity. At Sydney, in Australia, a single colony is stated to have multiplied to 300 in three years. All the new swarms except the first, are led off by a young queen, and as she is never impregnated until after she has been established as the head of a separate family, it is important that they should all be accompanied by a goodly number of drones; and this renders it necessary that a large number should be produced in the parent hive.

As this necessity no longer exists, when the bee is domesticated, the production of so many drones should be discouraged. Traps have been invented to destroy them, but it is much better to save the bees the labor and expense of rearing such a host of useless consumers. This can readily be done by the use of my hives. The cells in which the drones are reared, are much larger than those appropriated to the raising of workers. The combs containing them may be taken out, to have their places supplied with worker's cells, and thus the over production of drones may easily be prevented. Some colonies contain so much drone comb as to be nearly worthless.

I have no doubt that some of my readers will object to this mode of management as interfering with nature: but let them remember that the bee is not in a state of nature, and that the same objection might be urged against killing off the super-numerary males of our domestic animals.

In July or August, soon after the swarming season is over, the bees expel the drones from the hive. They sometimes sting them, and sometimes gnaw the roots of their wings, so that when driven from the hive, they cannot return. If not treated in either of these summary ways, they are so persecuted and starved, that they soon perish. The hatred of the bees extends even to the young which are still unhatched: they are mercilessly pulled from the cells, and destroyed with the rest. How wonderful that instinct which teaches the bees that there is no longer any occasion for the services of the drones, and which impels them to destroy those members of the colony, which, a short time before, they reared with such devoted attention!

A colony which neglects to expel its drones at the usual season, ought always to be examined. The queen is probably either diseased or dead. In my hives, such an examination may be easily made, the true state of the case ascertained, and the proper remedies at once applied. (See Chapter on the Loss of the Queen.)

The Production of so many Drones Necessary, in a State of Nature, to Prevent Degeneracy from "In and In Breeding."

I have often been able, by the reasons previously assigned, to account for the necessity of such a large number of drones in a state of nature, to the satisfaction of others, but never fully to my own. I have repeatedly queried, why impregnation might not just as well have been effected *in the hive*, as on the wing, in the open air. Two very obvious and highly important advantages would have resulted from such an arrangement. 1st. A few dozen drones would have amply sufficed for the wants of any colony, even if, (as in tropical climates,) it swarmed half a dozen times or oftener, in the same season. 2d. The young queens would have been exposed to none of those risks which they now incur, in leaving the hive for fecundation.

I was unable to show how the existing arrangement is best; although I never doubted that there must be a satisfactory reason for this seeming imperfection. To suppose otherwise, would be highly unphilosophical, since we constantly see, as the circle of our knowledge is enlarged, many mysteries in nature hitherto inexplicable, fully cleared up.

Let me here ask if the disposition which too many students of nature cherish, to reject some of the doctrines of revealed religion, is not equally unphilosophical. Neither our ignorance of all the facts necessary to their full elucidation, nor our inability to harmonize these facts in their mutual relations and dependencies, will justify us in rejecting any truth which God has seen fit to reveal, either in the book of nature, or in His holy word. The man who would substitute his own speculations for the divine teachings, has embarked, without rudder or chart, pilot or compass, upon the uncertain ocean of theory and conjecture; unless he turns his prow from its fatal course, no Sun of Righteousness will ever brighten for him the dreary expanse of waters; storms and whirlwinds will thicken in gloom, on his "voyage of life," and no favoring gales will ever waft his shattered bark to a peaceful haven.

The thoughtful reader will require no apology for the moralizing strain of many of my remarks, nor blame a clergyman, if forgetting sometimes to speak as the mere naturalist, he endeavors to find,

"Tongues in trees, books in the running brooks, *Sermons* in '*bees*,' and 'God' in every thing."

To return to the point from which I have digressed; a new attempt to account for the existence of so many drones. If a farmer persists in what is called "breeding in and in," that is, from the same stock without changing the blood, it is well known that a rapid degeneracy is the inevitable consequence. This law extends, as far as we know, to all animal life, and even man is not exempt from its influence. Have we any reason to suppose that the bee is an exception? or that ultimate degeneracy would not ensue, unless some provision was made to counteract the tendency to in and in breeding? If fecundation had taken place in the hive, the queen bee must of necessity, have been impregnated by drones from a common parent, and the same result must have taken place in each successive generation, until the whole species would eventually have "run out." By the present arrangement, the young females, when they leave the hive, often find the air swarming with drones, many of which belong to other colonies, and thus by crossing the breed, a provision is constantly made to prevent deterioration.

Experience has proved not only that it is unnecessary to impregnation that there should be drones in the colony of the young queen, but that this may be effected even when there are no drones in the Apiary, and none except at some considerable distance. Intercourse takes place very high in the air, (perhaps that less risk may be incurred from birds,) and this is the more favorable to the continual crossing of stocks.

I am strongly persuaded that the decay of many flourishing stocks, even when managed with great care, is to be attributed to the fact that they have become enfeebled by "close breeding," and are thus unable to resist the injurious influences which were comparatively harmless when the bees were in a state of high physical vigor. I shall, in the chapter on Artificial Swarming, explain in what way, by the use of my hives, the stock of bees may be easily crossed, when a cultivator is too remote from other Apiaries, to depend upon its being naturally effected.

THE WORKERS OR COMMON BEES.

The number of workers in a hive varies very much. A good swarm ought to contain 15,000 or 20,000; and in large hives, strong colonies which are not reduced by swarming, frequently number two or three times as many, during the height of the breeding season. We have well-authenticated instances of stocks much more populous than this. The Polish hives will hold several bushels, and yet we are informed by Mr. Dohiogost, that they swarm regularly, and that the swarms are so powerful that "they resemble a little cloud in the air." I shall hereafter consider how the size of the hive affects the number of bees that it may be expected to produce.

The workers, (as has been already stated,) are all females whose ovaries are too imperfectly developed to admit of their laying eggs. For a long time, they were regarded as neither males nor females, and were called Neuters; but more careful microscopic examinations have enabled us to detect the rudiments of their ovaries, and thus to determine their sex. The accuracy of these examinations has been verified by the well-known facts respecting *fertile workers*.

Riem, a German Apiarian, first discovered that workers sometimes lay eggs. Huber, in the course of his investigations on this subject, ascertained that such workers were raised in hives that had lost their queen, and in the vicinity of the royal cells in which young queens were being reared. He conjectured that they received accidentally, a small portion of the peculiar food of these infant queens, and in this way, he accounted for their reproductive organs being more developed than those of other workers. Workers reared in such hives, are in close proximity to the young queens, and there is certainly much probability that some of the royal jelly may be accidentally dropped into their cells; as, in these hives, the queen cells when first commenced are parallel to the horizon, instead of being perpendicular to it, as they are in other hives. I do not feel confident, however, that they are not sometimes bred in hives which have not lost their queen. The kind of eggs laid by these fertile workers, has already been noticed. Such workers are seldom tolerated in hives containing a fertile, healthy queen, though instances of this kind have been known to occur. The worker is much smaller than either the queen or the drone.[5] It is furnished with a tongue or proboscis, of the most curious and complicated structure, which, when not in use, is nicely folded under its abdomen; with this, it licks or brushes up the honey, which is thence conveyed to its honey-bag. This receptacle is not larger than a very small pea, and is so perfectly transparent, as to appear when filled, of the same color with its contents; it is properly the first stomach of the bee, and is surrounded by muscles which enable the bee to compress it, and empty its contents through her proboscis into the cells. (See Chapter on Honey.)

The hinder legs of the worker are furnished with a spoon-shaped hollow or basket, to receive the pollen or bee bread which she gathers from the flowers. (See Chapter on Pollen.)

Every worker is armed with a formidable sting, and when provoked, makes instant and effectual use of her natural weapon. The sting, when subjected to microscopic examination, exhibits a very curious and complicated mechanism. "It is moved[6] by muscles which, though invisible to the eye, are yet strong enough to force the sting, to the depth of one twelfth of an inch, through the thick skin of a man's hand. At its root are situated two glands by which the poison is secreted: these glands uniting in one duct, eject the venemous liquid along the groove, formed by the junction of the two piercers. There are four barbs on the outside of each piercer: when the insect is prepared to sting, one of these piercers, having its point a little longer than the other, first darts into the flesh, and being fixed by its foremost beard, the other strikes in also, and they alternately penetrate deeper and deeper, till they acquire a firm hold of the flesh with their barbed hooks, and then follows the sheath, conveying the poison into the wound. The action of the sting, says Paley, affords an example of the union of *chemistry* and mechanism; of chemistry in respect to the *venom*, which can produce such powerful effects; of mechanism as the sting is a compound instrument. The machinery would have been comparatively useless had it not been for the chemical process, by which in the insect's body *honey* is converted into *poison*; and on the other hand, the poison would have been ineffectual, without an instrument to wound, and a syringe to inject it."

"Upon examining the edge of a very keen razor by the microscope, it appears as broad as the back of a pretty thick knife, rough, uneven, and full of notches and furrows, and so far from anything like sharpness, that an instrument, as blunt as this seemed to be, would not serve even to cleave wood. An exceedingly small needle being also examined, it resembled a rough iron bar out of a smith's forge. The sting of a bee viewed through the same instrument, showed everywhere a polish amazingly beautiful, without the least flaw, blemish, or inequality, and ended in a point too fine to be discerned."

The extremity of the sting being barbed like an arrow, the bee can seldom withdraw it, if the substance into which she darts it is at all tenacious. In losing her sting she parts with a portion of her intestines, and of necessity, soon perishes.

As the loss of the sting is always fatal to the bees, they pay a dear penalty for the exercise of their patriotic instincts; but they always seem ready, (except when they have taken "a drop too much," and are gorged with honey,) to die in defence of their home and treasures; or as the poet has expressed it, they

"Deem life itself to vengeance well resign'd, Die on the wound, and leave their sting behind."

Hornets, wasps and other stinging insects are able to withdraw their stings from the wound. I have never seen any attempt to account for the exception in the case of the honey bee. But if the Creator intended the bee for the use of man, as He most certainly did, has He not given it this peculiarity, to make it less formidable, and therefore more completely subject to human control? Without a sting, it would have stood no chance of defending its tempting sweets against a host of greedy depredators; but if it could sting a number of times, it would be much more difficult to bring it into a state of thorough domestication. A quiver full of arrows in the hand of a skilful marksman, is far more to be dreaded than a single shaft.

The defence of the colony against enemies, the construction of the cells, the storing of them with honey and bee-bread, the rearing of the young, in short, the whole work of the hive, the laying of eggs excepted, is carried on by the industrious little workers.

There may be *gentlemen* of leisure in the commonwealth of bees, but most assuredly there are no such *ladies*, whether of high or low degree. The queen herself, has her full share of duties, for it must be admitted that the royal office is no sinecure, when the mother who fills it, must superintend daily the proper deposition of several thousand eggs!

AGE OF BEES.

The queen bee, (as has been already stated,) will live four, and sometimes, though very rarely, five years. As the life of the drones is usually cut short by violence, it is not easy to ascertain its precise limit. Bevan, in some interesting statements on the longevity of bees, estimates it not to exceed four months. The workers are supposed by him, to live six or seven months. Their age depends, however, very much upon their greater or less exposure to injurious influences and severe labors. Those reared in the spring and early part of summer, and on whom the heaviest labors of the hive must necessarily devolve, do not appear to live more than two or three months, while those which are bred at the close of summer, and early in autumn, being able to spend a large part of their time in repose, attain a much greater age. It is very evident that "the bee," (to use the words of a quaint old writer,) "is a summer bird," and that with the exception of the queen, none live to be a year old.

Notched and ragged wings, instead of gray hairs and wrinkled faces, are the signs of old age in the bee, and indicate that its season of toil will soon be over. They appear to die rather suddenly, and often spend their last days, and sometimes even their last hours, in useful labors. Place yourself before a hive, and see the indefatigable energy of these aged veterans, toiling along with their heavy burdens, side by side with their more youthful compeers, and then say if you can, that *you* have done work enough, and that you will give yourself up to slothful indulgence, while the ability for useful labor still remains. Let the cheerful hum of their industrious old age inspire you with better resolutions, and teach you how much nobler it is to meet death in the path of duty, striving still, as you "have opportunity," to "do good unto all men."

The age which individual members of the community may attain, must not be confounded with that of the colony. Bees have been known to occupy the same domicile for a great number of years. I have seen flourishing colonies which were twenty years old, and the Abbe Della Rocca speaks of some over forty years old! Such cases have led to the erroneous opinion that bees are a long-lived race. But this, as Dr. Evans has observed, is just as wise as if a stranger, contemplating a populous city, and personally unacquainted with its inhabitants, should on paying it a second visit, many years afterwards, and finding it equally populous, imagine that it was peopled by the same individuals, not one of whom might then be living.

"Like leaves on trees, the race of bees is found, Now green in youth, now withering on the ground; Another race the Spring or Fall supplies, They droop successive, and successive rise."

The cocoons spun by the larvæ, are never removed by the bees; they stick so closely to the sides of the cells, that the knowing bee well understands that the labor of removal would cost more than it would be worth. In process of time, the breeding cells become too small for the proper development of the young. In some cases, the bees must take down and reconstruct the old combs, for if they did not, the young issuing from them would always be dwarfs; whereas I once compared with other bees, those of a colony more than fifteen years old, and found no perceptible difference. That they do not always renew the old combs, must be admitted, as the young from some old hives are often considerably below the average size. On this account, it is very desirable to be able to remove the old combs occasionally, that their place may be supplied with new ones.

It is a great mistake to imagine that the brood combs ought to be changed every year. In my hives, they might, if it were desirable, be easily changed several times in a year: but once in five or six years is often enough; oftener than this requires a needless consumption of honey to replace them, besides being for other reasons undesirable, as the bees are always in winter, colder in new comb than in old. Inventors of hives have too often been, most emphatically "men of one idea:" and that one, instead of being a well established and important fact in the physiology of the bee, has frequently, (like the necessity for a yearly change of the brood combs,) been merely a conceit, existing nowhere but in the brain of a visionary projector. This is all harmless enough, until an effort is made to impose such miserable crudities upon an ignorant public, either in the shape of a patented hive, *or worse still, of an UNPATENTED hive, the pretended RIGHT to use which, is FRAUDULENTLY sold to the cheated purchaser*!!

For want of proper knowledge with regard to the age of bees, huge "bee palaces," and large closets in garrets or attics, have been constructed, and their proprietors have vainly imagined that the bees would fill them, however roomy; for they can see no reason why a colony should not continue to increase indefinitely, until at length it numbers its inhabitants by millions or billions! As the bees can never at one time equal, still less exceed the number which the queen is capable of producing in one season, these spacious dwellings have always an abundance of "spare rooms." It seems strange that men can be thus deceived, when often in their own Apiary, they have healthy stocks which have not swarmed for a year or more, and which yet in the spring are not a whit more populous than those which have regularly parted with vigorous swarms.

It is certain that the Creator, has for some wise reason, set a limit to the increase of numbers in a single colony; and I shall venture to assign what appears to me to have been one reason for His so doing. Suppose that He had given to the bee, a length of life as great as that of the horse or the cow, or had made each queen capable of laying daily, some hundreds of thousands of eggs, or had given several hundred queens to each hive, then from the Very nature of the case, a colony must have gone on increasing, until it became a scourge rather than a benefit to man. In the warm climates of which the bee is a native, they would have established themselves in some cavern or capacious cleft in the rocks, and would there have quickly become so powerful as to bid defiance to all attempts to appropriate the avails of their labors.

It has already been stated, that none, except the mother wasps and hornets, survive the winter. If these insects had been able, like the bee, to commence the season with the accumulated strength of a large colony, long before its close, they would have proved a most intolerable nuisance. If, on the contrary, the queen bee had been compelled, solitary and alone, to lay the foundations of a new commonwealth, the honey-harvest would have disappeared before she could have become the parent of a numerous family.

In the laws which regulate the increase of bees as well as in all other parts of their economy, we have the plainest proofs that the insect was formed for the special service of the human race.

The process of rearing the Queen more particularly described.

If in the early part of the season, the population of a hive becomes uncomfortably crowded, the bees usually make preparations for swarming. A number of royal cells are commenced, and they are placed almost always upon those edges of the combs which are not attached to the sides of the hive. These cells somewhat resemble a small ground-nut or pea-nut, and are about an inch deep, and one-third of an inch in diameter: they are very thick, and require a large quantity of material for their construction. They are seldom seen in a perfect state, as the bees nibble them away after the queen has hatched, leaving only their remains, in the shape of a very small acorn-cup. While the other cells open sideways, these always hang with their mouth *downwards.* Much speculation has arisen as to the reason for this deviation: some have conjectured that their peculiar position exerted an influence upon the development of the royal larvæ; while others, having ascertained that no injurious effect was produced by turning them upwards, or placing them in any other position, have considered this deviation as among the inscrutable mysteries of the bee-hive. So it always seemed to me, until more careful reflection enabled me to solve the problem. The queen cells open downwards, simply *to save room!* The distance between the parallel ranges of comb being usually less than half an inch, the bees could not have made the royal cells to open sideways, without sacrificing the cells opposite to them. In order to economize space, to the very utmost, they put them upon the unoccupied edges of the comb, as the only place where there is always plenty of room for such very large cells.

The number of royal cells varies greatly; sometimes there are only two or three, ordinarily there are five or six, and I have occasionally seen more than a dozen. They are not all commenced at once, for the bees do not intend that the young queens shall all arrive at maturity, at the same time. I do not consider it as fully settled, how the eggs are deposited in these cells. In some few instances, I have known the bees to transfer the eggs from common to queen cells, and this *may* be their general method of procedure. I shall hazard the conjecture that the queen deposits her eggs in cells on the edges of the comb, in a crowded state of the hive, and that some of these are afterwards enlarged and changed into royal cells by the workers. Such is the instinctive hatred of the queen to her own kind, that it does not seem to me probable, that she is intrusted with even the initiatory steps for securing a race of successors. That the eggs from which the young queens are produced, are of the same kind with those producing workers, has been repeatedly demonstrated. On examining the queen cells while they are in progress, one of the first things which excites our notice, is the very unusual amount of attention bestowed upon them by the workers. There is scarcely a second in which a bee is not peeping into them, and just as fast as one is satisfied, another pops its head in, to examine if not to report, progress. The importance of their inmates to the bee-community, might easily be inferred from their being the center of so much attraction.

ROYAL JELLY.

The young queens are supplied with a much larger quantity of food than is allotted to the other larvæ, so that they seem almost to float in a thick bed of jelly, and there is usually a portion of it left unconsumed at the base of the cells, after the insects have arrived at maturity. It is different from the food of either drones or workers, and in appearance, resembles a light quince jelly, having a slightly acid taste.

I submitted a portion of the royal jelly for analysis, to Dr. Charles M. Wetherill, of Philadelphia; a very interesting account of his examination may be found in the proceedings of the Phila. Academy of Nat. Sciences for July, 1852. He speaks of the substance as "truly a bread-containing, albuminous compound." I hope in the course of the coming summer to obtain from this able analytical chemist, an analysis of the food of the young drones and workers. A comparison of its elements with those of the royal jelly, may throw some light on subjects as yet involved in obscurity.

The effects produced upon the larvæ by this peculiar food and method of treatment, are very remarkable. For one, I have never considered it strange that such effects should be rejected as idle whims, by nearly all except those who have either been eye-witnesses to them, or have been well acquainted with the character and opportunities for accurate observation, of those on whose testimony they have received them. They are not only in themselves most marvelously strange, but on the face of them so entirely opposed to all common analogies, and so very improbable, that many men when asked to believe them, feel almost as though an insult were offered to their common sense. The most important of these effects, I shall now proceed to enumerate.

1st. The peculiar mode in which the worm designed to be reared as a queen, is treated, causes it to arrive at maturity, about one-third earlier than if it had been bred a worker. And yet it is to be much more fully developed, and according to ordinary analogy, ought to have had a *slower growth!*

2d. Its organs of reproduction are completely developed, so that it is capable of fulfilling the office of a mother.

3d. Its size, shape and color are all greatly changed. Its lower jaws are shorter, its head rounder, and its legs have neither brushes nor baskets, while its sting is more curved, and one-third longer than that of a worker.

4th. Its *instincts* are entirely changed. Reared as a worker, it would have been ready to thrust out its sting, upon the least provocation; whereas now, it may be pulled limb from limb, without attempting to sting. As a worker it would have treated a queen with the greatest consideration; whereas now, if placed under a glass with another queen, it rushes forthwith to mortal combat with its rival. As a worker, it would frequently have left the hive, either for labor or exercise: as a queen, after impregnation, it never leaves the hive except to accompany a new swarm.

5th. The term of its life is remarkably lengthened. As a worker, it would have lived not more than six or seven months at farthest; as a queen it may live seven or eight times as long! All these wonders rest on the impregnable basis of complete demonstration, and instead of being witnessed by only a select few, may now, by the use of my hive, be familiar sights to any bee keeper, who prefers to acquaint himself with facts, rather than to cavil and sneer at the labors of others.[7]

When provision has been made, in the manner described, for a new race of queens, the old mother always departs with the first swarm, before her successors have arrived at maturity.[8]

ARTIFICIAL REARING OF QUEENS.

The distress of the bees when they lose their queen, has already been described. If they have the means of supplying her loss, they soon calm down, and commence forthwith, the necessary steps for rearing another. The process of rearing queens artificially, to meet some special emergency, is even more wonderful than the natural one, which has already been described. Its success depends on the bees having worker-eggs or worms not more than three days old; (if older, the larva has been too far developed as a worker to admit of any change:) the bees nibble away the partitions of two cells adjoining a third, so as to make one large cell out of the three. They destroy the eggs or worms in two of these cells, while they place before the occupant of the third, the usual food of the young queens, and build out its cell, so as to give it ample space for development. They do not confine themselves to the attempt to rear a single queen, but to guard against failure, start a considerable number, although the work on all except a few, is usually soon discontinued.

In twelve or fourteen days, they are in possession of a new queen, precisely similar to one reared in the natural way, while the eggs which were laid at the same time in the adjoining cells, and which have been developed in the usual way, are nearly a week longer in coming to maturity.

I will give in this connection a description of an interesting experiment:

A large hive which stood at a distance from any other colony, was removed in the morning of a very pleasant day, to a new place, and another hive containing only empty comb, was put upon its stand. Thousands of workers which were out in the fields, or which left the old hive after its removal, returned to the familiar spot. It was affecting to witness their grief and despair: they flew in restless circles about the place which once contained their happy home, entered and left the new hive continually, expressing, in various ways, their lamentations over their cruel bereavement. Towards evening, they ceased to take wing, and roamed in restless platoons, in and out of the hive, and over its surface, acting all the time, as though in search of some lost treasure. I now gave them a piece of brood comb, containing worker eggs and worms, taken from a second swarm which being just established with its young queen, in a new hive, could have no intention of rearing young queens that season; therefore, it cannot be contended that this piece of comb contained what some are pleased to call "royal eggs." What followed the introduction of this brood comb, took place much quicker than it can be described. The bees which first touched it, raised a peculiar note, and in a moment, the comb was covered with a dense mass; their restless motions and mournful noises ceased, and a cheerful hum at once attested their delight! Despair gave place to hope, as they recognized in this small piece of comb, the means of deliverance. Suppose a large building filled with thousands of persons, tearing their hair, beating their breasts, and by piteous cries, as well as frantic gestures, giving vent to their despair; if now some one should enter this house of mourning, and by a single word, cause all these demonstrations of agony to give place to smiles and congratulations, the change could not be more wonderful and instantaneous, than that produced when the bees received the brood comb!

The Orientals call the honey bee, Deburrah, "She that speaketh." Would that this little insect might speak, and in words more eloquent than those of man's device, to the multitudes who allow themselves to reject the doctrines of revealed religion, because, as they assert, they are, on their face so utterly improbable, that they labor under an *a priori* objection strong enough to be fatal to their credibility. Do not nearly all the steps in the development of a queen from a worker-egg, labor under precisely the same objection? and have they not, for this very reason, always been regarded by great numbers of bee keepers, as unworthy of credence? If the favorite argument of infidels and errorists will not stand the test when applied to the wonders of the bee-hive, can it be regarded as entitled to any serious weight, when employed in framing objections against religious truths, and arrogantly taking to task the infinite Jehovah, for what He has been pleased to do or to teach? Give me the same latitude claimed by such objectors, and I can easily prove that a man is under no obligation to receive any of the wonders in the economy of the bee-hive, although he is himself an intelligent eye-witness that they are all substantial verities.

I shall quote, in this connection, from Huish, an English Apiarian of whom I have already spoken, because his objections to the discoveries of Huber, remind me so forcibly of both the spirit and principles of the great majority of those who object to the doctrines of revealed religion.

"If an individual, with the view of acquiring some knowledge of the natural history of the bee, or of its management, consult the works of Bagster, Bevan, or any of the periodicals which casually treat upon the subject, will he not rise from the study of them with his mind surcharged with falsities and mystification? Will he not discover through the whole of them a servile acquiescence in the opinions and discoveries of one man, however at variance they may be with truth or probability; and if he enter upon the discussion with his mind free from prejudice, will he not experience that an outrage has been committed upon his reason, in calling upon him to give assent to positions and principles which at best are merely assumed, but to which he is called upon dogmatically to subscribe his acquiescence as the indubitable results of experience, skill and ability? The editors of the works above alluded to, should boldly and indignantly have declared, that from their own experience in the natural economy of the insect, they were able to pronounce the circumstances as related by Huber to be directly *impossible*, and the whole of them based on fiction and imposition."

Let the reader change only a few words in this extract: for "the natural history of the bee or its management," let him write, "the subject of religion;" for, "the works of Bagster, Bevan," &c., let him put, "the works of Moses, Paul," &c.; for, "their own experience in the natural economy of the insect," let him substitute, "their own experience in the nature of man;" and for, "circumstances as related by Huber," let him insert, "as related by Luke or John," and it will sound almost precisely like a passage from some infidel author.

I resume the quotation from Huish; "If we examine the account which Huber gives of his invention (!) of the royal jelly, the existence and efficacy of which are fully acquiesced in by the aforesaid editors, to what other conclusions are we necessarily driven, than that they are the dupes of a visionary enthusiast, whose greatest merit consists in his inventive powers, no matter how destitute those powers may be of all affinity with truth or probability? Before, however, these editors bestowed their unqualified assent on the existence of this royal jelly, did they stop to put to themselves the following questions? By what kind of bee is it made?[9] Whence is it procured? Is it a natural or an elaborated substance? If natural, from what source is it derived? If elaborated, in what stomach of the bee is it to be found? How is it administered? What are its constituent principles? Is its existence optional or definite? Whence does it derive its miraculous power of converting a common egg into a royal one? Will any of the aforesaid editors publicly answer these questions? and ought they not to have been able to answer them, before they so unequivocally expressed their belief in its existence, its powers and administration?"

How puerile does all this sound to one who has *seen* and *tasted* the royal jelly! And permit me to add, how equally unmeaning do the objections of infidels seem, to those who have an experimental acquaintance with the divine hopes and consolations of the Gospel of Christ.

CHAPTER IV. COMB.

W ax is a natural secretion of the bees; it may be called *their oil or fat*. If they are gorged with honey, or any liquid sweet, and remain quietly clustered together, it is formed in small wax pouches on their abdomen, and comes out in the shape of very delicate scales. Soon after a swarm is hived, the bottom board will be covered with these scales.

"Thus, filtered through yon flutterer's folded mail, Clings the cooled wax, and hardens to a scale. Swift, at the well known call, the ready train, (For not a buz boon Nature breathes in vain,) Spring to each falling flake, and bear along Their glossy burdens to the builder throng. These with sharp sickle or with sharper tooth, Pare each excrescence, and each angle smooth, Till now, in finish'd pride, two radiant rows Of snow white cells one mutual base disclose. Six shining panels gird each polish'd round, The door's fine rim, with waxen fillet bound, While walls so thin, with sister walls combined, Weak in themselves, a sure dependence find." *Evans*.

Huber was the first to demonstrate that wax is a natural secretion of the bee, when fed on honey or any saccharine substance. Most Apiarians before his time, supposed that it was made from pollen or bee-bread, either in a crude or digested state. He confined a new swarm of bees in a hive placed in a dark and cool room, and on examining them, at the end of five days, found several beautiful white combs in their tenement: these were taken from them, and they were again confined and supplied with honey and water, and a second time new combs were constructed. Five times in succession their combs were removed, and were in each instance replaced, the bees being all the time prevented from ranging the fields, to supply themselves with bee-bread. By subsequent experiments he proved that sugar answered the same end with honey.

He then confined a swarm, giving them no honey, but an abundance of fruit and pollen. They subsisted on the fruit, but refused to touch the pollen; and no combs were constructed, nor any wax scales formed in their pouches. These experiments are conclusive; and are interesting, not merely as proving that wax is secreted from honey or saccharine substances, but because they show in what a thorough manner the experiments of Huber were conducted. Confident assertions are easily made, requiring only a little breath or a drop of ink; and the men who deal most in them, have often a profound contempt for observation and experiment. To establish even a simple truth, on the solid foundation of demonstrated facts, often requires the most patient and protracted toil.

A high temperature is necessary for comb-building, in order that the wax may be soft enough to be moulded into shape. The very process of its secretion helps to furnish the amount of heat which is required to work it. This is a very interesting fact which seems never before to have been noticed.

Honey or sugar is found to contain by weight, about eight pounds of oxygen to one of carbon and hydrogen. When changed into wax, the proportions are entirely reversed: the wax contains only one pound of oxygen to more than sixteen pounds of hydrogen and carbon. Now as oxygen is the grand supporter of animal heat, the consumption of so large a quantity of it, aids in producing the extraordinary heat which always accompanies comb-building, and which is necessary to keep the wax in the soft and plastic state requisite to enable the bees to mould it into such exquisitely delicate and beautiful shapes! Who can fail to admire the wisdom of the Creator in this beautiful instance of adaptation?

The most careful experiments have clearly established the fact, that at least *twenty pounds* of honey are consumed in making a single pound of wax. If any think that this is incredible, let them bear in mind that wax is an animal oil secreted from honey, and let them consider how many pounds of corn or hay they must feed to their stock, in order to have them gain a single pound of fat.

Many Apiarians are entirely ignorant of the great value of empty comb. Suppose the honey to be worth only 15 cts. per lb., and the comb when rendered into wax, to be worth 30 cts. per lb., the bee-master who melts a pound of comb, loses nearly three dollars by the operation, and this, without estimating the time which the bees have consumed in building the comb. Unfortunately, in the ordinary hives, but little use can be made of empty comb, unless it is new, and can be put into the surplus honey-boxes: but by the use of my movable frames, every piece of good worker-comb may be used to the best advantage, as it can be given to the bees, to aid them in their labors.

It has been found very difficult to preserve comb from the bee-moth, when it is taken from the bees. If it contains only a *few* of the eggs of this destroyer, these, in due time, will produce a progeny sufficient to devour it. The comb, if it is attached to my frames, may be suspended in a box or empty hive, and thoroughly smoked with sulphur; this will kill any *worms* which it may contain. When the weather is warm enough to hatch the eggs of the moth, this process must be repeated a few times, at intervals of about a week, so as to insure the destruction of the worms as they hatch, for the sulphur does not seem always to destroy the vitality of the eggs. The combs may now be kept in a tight box or hive, with perfect safety.

Combs containing bee-bread, are of great value, and if given to young colonies, which in spring are frequently destitute of this article, they will materially assist them in early breeding.

Honey may be taken from my hives in the frames, and the covers of the cells sliced off with a sharp knife; the honey can then be drained out, and the empty combs returned to be filled again. A strong stock of bees, in the height of the honey harvest, will fill empty combs with wonderful rapidity. I lay it down, as one of my *first principles* in bee culture, that no good comb should ever be melted; it should all be carefully preserved and given to the bees. If it is new, it may be easily attached to the frames, or the honey-receptacles, by dipping the edge into melted wax, pressing it gently until it stiffens, and then allowing it to cool. If the comb is old, or the pieces large and full of bee-bread, it will be best to dip them into melted rosin, which, besides costing much less than wax, will secure a much firmer adhesion. When comb is put into tumblers or other small vessels, the bees will begin to work upon it the sooner, if it is simply crowded in, so as to be held in place by being supported against the sides. It would seem as though they were disgusted with such unworkmanlike proceedings, and that they cannot rest until they have taken it into hand, and endeavored to "make a job of it."

If the bee-keeper in using his choicest honey will be satisfied to dispense with looks, and will carefully drain it from the beautiful comb, he may use all such comb again to great advantage; not only saving its intrinsic value, but greatly encouraging his bees to occupy and fill all receptacles in which a portion of it is put. Bees seem to fancy *a good start in life*, about as well as their more intelligent owners. To this use all suitable drone comb should be put, as soon as it is removed from the main hive. (See remarks on Drones.)

Ingenious efforts have been made, of late years, to construct *artificial* honey combs of porcelain, to be used for *feeding* bees. No one, to my knowledge, has ever attempted to imitate the delicate mechanism of the bee so closely, as to construct artificial combs for the ordinary uses of the hive; although for a long time I have entertained the idea as very desirable, and yet as barely possible. I am at present engaged in a course of experiments on this subject, the results of which, in due time, I shall communicate to the public.

While writing this treatise, it has occurred to me that bees might be induced to use old wax for the construction of their combs. Very fine parings may be shaved off with glass, and if given to the bees, under favorable circumstances, it seems to me very probable that they would use them, just as they do the scales which are formed in their wax pouches. Let strong colonies be deprived of some of their combs, after the honey harvest is over, and supplied abundantly with these parings of wax. Whether "nature abhors a vacuum," or not, bees certainly do, when it occurs among the combs of their main hive. They will not use the honey stored up for winter use to replace the combs taken from them; they can gather none from the flowers; and I have strong hopes that necessity will with bees as well as men, prove the mother of invention, and lead them to use the wax, as readily as they do the substitutes offered them for pollen. (See Chapter on Pollen.)

If this conjecture should be verified by actual results, it would exert a most powerful influence in the cheap and rapid multiplication of colonies, and would enable the bees to store up most prodigious quantities of honey. A pound of bees wax might then be made to store up twenty pounds of honey, and the gain to the bee keeper would be the difference in price between the pound of wax, and the twenty pounds of honey, which the bees would have consumed in making the same amount of comb. Strong stocks might thus during the dull season, when no honey can be procured, be most profitably employed in building spare comb, to be used in strengthening feeble stocks, and for a great variety of purposes. Give me the means of cheaply obtaining large amounts of comb, and I have almost found the philosopher's stone in bee keeping.

The building of comb is carried on with the greatest activity in the night, while the honey is gathered by day. Thus no time is lost. If the weather is too forbidding to allow the bees to go abroad, the combs are very rapidly constructed, as the labor is carried on both by day and by night. On the return of a fair day, the bees gather unusual quantities of honey, as they have plenty of room for its storage. Thus it often happens, that by their wise economy of time, they actually lose nothing, even if confined, for several days, to their hive.

"How doth the little busy bee, improve each *shining* hour!"

The poet might with equal truth have described her, as improving the gloomy days, and the dark nights, in her useful labors.

It is an interesting fact, which I do not remember ever to have seen particularly noticed by any writer, that honey gathering, and comb building, go on simultaneously; so that when one stops, the other ceases also. I have repeatedly observed, that as soon as the honey harvest fails, the bees intermit their labors in building new comb, even when large portions of their hive are unfilled. They might enlarge their combs by using some of their stores; but then they would incur the risk of perishing in the winter, by starvation. When honey no longer abounds in the fields, it is wisely ordered, that they should not consume their hoarded treasures, in expectation of further supplies, which may never come. I do not believe, that any other safe rule could have been given them; and if honey gathering was our business, with all our boasted reason, we should be obliged to adopt the very same course.

Wax is one of the best non-conductors of heat, so that when it is warmed by the animal heat of the bees, it can more easily be worked, than if it parted with its heat too readily. By this property, the combs serve also to keep the bees warm, and there is not so much risk of the honey candying in the cells, or the combs cracking with frost. If wax was a good conductor of heat, the combs would often be icy cold, moisture would condense and freeze upon them, and they would fail to answer the ends for which they are intended.

The size of the cells, in which workers are reared, never varies: the same may substantially be said of the drone cells which are very considerably larger; the cells in which honey is stored, often vary exceedingly in depth, while in diameter, they are of all sizes from that of the worker cells to that of the drones.

The cells of the bees are found perfectly to answer all the most refined conditions of a very intricate mathematical problem! Let it be required to find what shape a given quantity of matter must take, in order to have *the greatest capacity, and the greatest strength*, requiring at the same time, *the least space, and the least labor* in its construction. This problem has been solved by the most refined processes of the higher mathematics, and the result is the hexagonal or six-sided cell of the honey bee, with its three four-sided figures at the base!

The shape of these figures cannot be altered, *ever so little, except for the worse*. Besides possessing the desirable qualities already described, they answer as *nurseries* for the rearing of the young, and as *small air-tight vessels* in which the honey is preserved from souring or candying. Every prudent housewife who puts up her preserves in tumblers, or small glass jars, and carefully pastes them over, to keep out the air, will understand the value of such an arrangement.

"There are only three possible figures of the cells," says Dr. Reid, "which can make them all equal and similar, without any useless spaces between them. These are the equilateral triangle, the square and the regular hexagon. It is well known to mathematicians that there is not a fourth way possible, in which a plane may be cut into little spaces that shall be equal, similar and regular, without leaving any interstices."

An equilateral triangle would have made an uncomfortable tenement for an insect with a round body; and a square would not have been much better. At first sight a circle would seem to be the best shape for the development of the larvæ: but such a figure would have caused a needless sacrifice of space, materials and strength; while the honey which now adheres so admirably to the many angles or corners of the six-sided cell, would have been much more liable to run out! I will venture to assign a new reason for the hexagonal form. The body of the immature insect as it undergoes its changes, is charged with a super-abundance of moisture which passes off through the reticulated cover which the bees build over its cell: a hexagon while it approaches so nearly the shape of a circle as not to incommode the young bee, furnishes in its six corners the necessary vacancies for its more thorough ventilation!

So invariably uniform in size, as well as perfect in other respects, are the cells in which the workers are bred, that some mathematicians have proposed their adoption, as the best unit for measures of capacity to serve for universal use.

Can we believe that these little insects unite so many requisites in the construction of their cells, either by chance, or because they are profoundly versed in the most intricate mathematics? Are we not compelled to acknowledge that the mathematics must be referred to the Creator, and not to His puny creature? To an intelligent, candid mind, a piece of honey comb is a complete demonstration that there is a "GREAT FIRST CAUSE:" for on no other supposition can we account for so complicated a shape, and yet the only one which can possibly unite so many desirable requisites.

"On books deep poring, ye pale sons of toil, Who waste in studious trance the midnight oil, Say, can ye emulate with all your rules, Drawn or from Grecian or from Gothic schools, This artless frame? Instinct her simple guide, A heaven-taught Insect baffles all your pride. Not all yon marshall'd orbs, that ride so high, Proclaim more loud a present Deity, Than the nice symmetry of these small cells, Where on each angle genuine science dwells." *Evans.*

CHAPTER V. PROPOLIS, OR "BEE-GLUE."

This substance is obtained by the bees from the resinous buds and limbs of trees; and when first gathered, it is usually of a bright golden color, and is exceedingly sticky. The different kinds of poplars furnish a rich supply. The bees bring it on their thighs just as they do bee bread; and I have caught them as they were entering with a load, and taken it from them. It adheres so firmly that it is difficult to remove it.

"Huber planted in Spring some branches of the wild poplar, before the leaves were developed, and placed them in pots near his Apiary; the bees alighting on them, separated the folds of the largest buds with their forceps, extracted the varnish in threads, and loaded with it, first one thigh and then the other; for they convey it like pollen, transferring it by the first pair of legs to the second, by which it is lodged in the hollow of the third." The smell of the propolis is often precisely similar to that of the resin from the poplar, and chemical analysis proves the identity of the two substances. It is frequently gathered from the alder, horse-chestnut, birch, and willow; and as some think, from pines and other trees of the fir kind. I have often known bees to enter the shops where varnishing was being carried on, attracted evidently by the smell: and Bevan mentions the fact of their carrying off a composition of wax and turpentine, from trees to which it had been applied. Dr. Evans says that he has seen them collect the balsamic varnish which coats the young blossom buds of the hollyhock, and has known them to rest at least ten minutes on the same bud, moulding the balsam with their fore feet, and transferring it to the hinder legs, as described by Huber.

"With merry hum the Willow's copse they scale, The Fir's dark pyramid, or Poplar pale, Scoop from the Aider's leaf its oozy flood, Or strip the Chestnut's resin-coated bud, Skim the light tear that tips Narcissus' ray, Or round the Hollyhock's hoar fragrance play. Soon temper'd to their will through eve's low beam, And link'd in airy bands the viscous stream, They waft their nut-brown loads exulting home, That form a fret-work for the future comb; Caulk every chink where rushing winds may roar, And seal their circling ramparts to the floor." *Evans.*

A mixture of wax and propolis is used by the bees to strengthen the attachments of the combs to the top and sides of the hive, and serves most admirably for this purpose, as it is much more adhesive than wax alone. If the combs, as soon as they are built, are not filled with honey or brood, they are beautifully varnished with a most delicate coating of this material, which adds exceedingly to their strength: but as this natural varnish impairs their delicate whiteness, they ought not to be allowed to remain in the surplus honey receptacles, accessible to the bees, unless when they are actively engaged in storing them with honey.

The bees make a very liberal use of this substance to fill up all the crevices about their premises: and as the natural summer heat of the hive keeps it soft, the bee moth selects it as a proper place of deposit for her eggs. For this reason, the hive should be made of sound lumber, entirely free from cracks, and thoroughly painted on the inside as well as outside. When glass is used, there is no risk that the bed moth will find a place in which she can insert her ovi-positor and lay her eggs. The corners of the hive, which the bees always fill with propolis, should have a melted mixture of three parts rosin, and one part bees-wax run into them, which remains hard during the hottest weather, and bids defiance to the moth. The inside of the hive may be coated with the same mixture, put on hot with a brush.

The bees find it difficult to gather the propolis, and equally so to remove from their thighs, and to work so sticky a material. For this reason, it is doubly important to save them all unnecessary labor in amassing it. To men, time is *money*; to bees, it is *honey*; and all the arrangements of the hive should be such as to economize it to the very utmost.

Propolis is sometimes put to a very curious use by the bees. "A snail[10] having crept into one of M. Reaumur's hives early in the morning, after crawling about for some time, adhered by means of its own slime to one of the glass panes. The bees having discovered the snail, surrounded it and formed a border of propolis round the verge of its shell, and fastened it so securely to the glass that it became immovable."

"Forever closed the impenetrable door, It naught avails that in his torpid veins Year after year, life's loitering spark remains."[11] *Evans.*

"Maraldi, another eminent Apiarian, has related a somewhat similar instance. He states that a snail without a shell, or slug, as it is called, had entered one of his hives; and that the bees, as soon as they observed it, stung it to death: after which being unable to dislodge it, they covered it all over with an impervious coat of propolis."

"For soon in fearless ire, their wonder lost, Spring fiercely from the comb the indignant host, Lay the pierced monster breathless on the ground, And clap in joy their victor pinions round: While all in vain concurrent numbers strive, To heave the slime-girt giant from the hive— Sure not alone by force Instinctive swayed, But blest with reason's soul directing aid, Alike in man or bee, they haste to pour, Thick hard'ning as it falls, the flaky shower; Embalmed in shroud of glue the mummy lies, No worms invade, no foul miasmas rise." *Evans.*

"In these cases who can withhold his admiration of the ingenuity and judgment of the bees? *In the first case* a troublesome creature gained admission to the hive, which, from its unwieldiness, they could not remove, and which, from the impenetrability of its shell, they could not destroy: here then their only resource was to deprive it of locomotion, and to obviate putrefaction; both which objects they accomplished most skilfully and securely—and as is usual with these sagacious creatures, at the least possible expense of labor and materials. They applied their cement where alone it was required, round the verge of the shell. *In the latter case*, to obviate the evil of decay, by the total exclusion of air, they were obliged to be more lavish in the use of their embalming material, and to case over the "slime girt giant" so as to guard themselves from his noisome smell. What means more effectual could human wisdom have devised under similar circumstances?"

"If in the insect, Season's twilight ray Sheds on the darkling mind a doubtful day, Plain is the steady light her *Instincts* yield, To point the road o'er life's unvaried field; If few these instincts, to the destined goal, With surer coarse, their straiten'd currents roll." *Evans.*

CHAPTER VI. POLLEN, OR BEE-BREAD.

This substance is gathered by the bees from the flowers, or blossoms, and is used *for the nourishment of their young*. Repeated experiments have proved that no brood can be raised in a hive, unless the bees are supplied with it. It contains none of the elements of wax, but is rich in what chemists call nitrogenous substances, which are not contained in honey, and which furnish ample nourishment for the development of the growing bee. Dr. Hunter dissected some immature bees, and found their stomachs to contain farina, but not a particle of honey.

We are indebted to Huber for the discovery of the use made by the bees of pollen. That it did not serve as food for the mature bees, was evident from the fact that large supplies are often found in hives whose inmates have starved to death. It was this fact which led the old observers to conclude that it was gathered for the purpose of building comb. After Huber had demonstrated that wax is secreted from an entirely different substance, he was soon led to conjecture that the bee-bread must be used for the nourishment of the embryo bees. By rigid experiments he proved the truth of this supposition. Bees were confined to their hive without any pollen, after being supplied with honey, eggs and larvæ. In a short time the young all perished. A fresh supply of brood was given to them, with an ample allowance of pollen, and the development of the larvæ then proceeded in the natural way.

When a colony is actively engaged in carrying in this article, it may be taken for granted that they have a fertile queen, and are busy in breeding. On the contrary, if any colony is not gathering pollen when others are, the queen is either dead, or diseased, and the hive should at once be examined.

In the backward spring of 1852, I had an excellent opportunity of testing the value of this substance. In one of my hives, was an artificial swarm of the previous year. The hive was well protected, being double, and the situation was warm. I opened it on the 5th of February, and although the weather, until within a week of that time, had been unusually cold, I found many of the cells filled with brood. On the 23d, the combs were again examined, and found to contain, neither eggs, brood, nor bee bread. The bees were then supplied with bee bread taken from another hive: the next day, this was found to have been used by them, and a large number of eggs had been deposited in the cells. When this supply was exhausted, egg-laying ceased, and was again renewed when more was furnished them.

During all the time of these experiments, the weather was unpromising, and as the bees were unable to go out for water, they were supplied at home with this important article.

Dzierzon is of opinion that the bees are able to furnish food for the young, without the presence of pollen in the hive; although he admits that they can do this only for a short time, and at a great expense of vital energy; just as the strength of an animal nursing its young is rapidly reduced, when for want of proper food, the very substance of its own body as it were, is converted into milk. My experiments do not corroborate this theory, but tend to confirm the views of Huber, and to show the absolute necessity of pollen to the development of brood. The same able contributor to Apiarian science, thinks that pollen is used by the bees when they are engaged in comb-building; and that unless they are well supplied with it, they cannot rapidly secrete wax, without very severely taxing their strength. But as all the elements of wax are found in honey, and none of them in pollen, this opinion does not seem to me, to be entitled to much weight. That bees cannot live upon pollen without any honey, is proved by the fact, that large stores of it are often found, in hives whose occupants have died of starvation; that they can live without it, is equally well known; but that the full grown bees make some use of it in connection with honey, for their own nourishment, I believe to be highly probable.

The bees prefer to gather *fresh* bee-bread, even when there are large accumulations of old stores in the cells. Hence, the great importance of being able to make the *surplus* of old colonies supply the *deficiency* of young ones. (See No. 28, in the Chapter "On the advantages which ought to be found in an Improved Hive.")

If both honey and pollen can be obtained from the same flower, then a load of *each* will be secured by the industrious insect. Of this, any one may convince himself, who will dissect a few pollen gatherers at the time when honey is plenty: he will generally find their honey-bags full.

The mode of gathering is very interesting. The body of the bee appears, to the naked eye, to be covered with fine hairs; to these, when the bee alights on a flower, the farina adheres. With her legs, she brushes it off from her body, and packs it in two hollows or *baskets*, one on each of her thighs: these baskets are surrounded by stouter hairs which hold the load in its place.

When the bee returns with pollen, she often makes a singular, dancing or vibratory motion, which attracts the attention of the other bees, who at once nibble away from her thighs what they want for immediate use; the rest she deposits in a cell for future need, where it is carefully packed down, and often sealed over with wax.

It has been observed that a bee, in gathering pollen, always confines herself to the same kind of flower on which she begins, even when that is not so abundant as some others. Thus if you examine a ball of this substance taken from her thigh, it is found to be of one uniform color throughout: the load of one will be yellow, another red, and a third brown; the color varying according to that of the plant from which it was obtained. It is probable that the pollen of different kinds of flowers would not pack so well together. It is certain that if they flew from one species to another, there would be a much greater mixture of different varieties than there now is, for they carry on their bodies the pollen or fertilizing principle, and thus aid most powerfully in the impregnation of plants.

This is one reason why it is so difficult to preserve pure, the different varieties of the same vegetables whose flowers are sought by the bee.

He must be blind indeed, who will not see, at every step in the natural history of this insect, the plainest proofs of the wisdom of its Creator.

I cannot resist the impression that the honey bee was made for the especial service and instruction of man. At first the importance of its products, when honey was the only natural sweet, served most powerfully to attract his attention to its curious habits; and now since the cultivation of the sugar cane has diminished the relative value of its luscious sweets, the superior knowledge which has been obtained of its instincts, is awakening an increasing enthusiasm in its cultivation.

Virgil in the fourth book of his Georgics, which is entirely devoted to bees, speaks of them as having received a direct emanation from the Divine Intelligence. And many modern Apiarians are almost disposed to rank the bee for sagacity, as next in the scale of creation to man.

The importance of pollen to the nourishment of the brood, has long been known, and of late, successful attempts have been made to furnish a *substitute*. The bees in Dzierzon's Apiary were observed by him, early in the spring before the time for procuring pollen, to bring rye meal to their hives from a neighboring mill. It is now a common practice on the continent of Europe, where bee keeping is extensively carried on, to supply the bees, in early spring, with this article. Shallow troughs are set in front of the Apiaries, which are filled, about two inches deep, with *finely ground, dry, unbolted rye meal.* Thousands of bees resort eagerly to them when the weather is favorable, roll themselves in the meal, and return heavily laden to their hives. In fine, mild weather, they labor at this work with astonishing industry; and seem decidedly to prefer the meal to the *old* pollen stored in their combs. By this means, the bees are induced to commence breeding *early*, and rapidly recruit their numbers. The feeding is continued till the bees cease to carry away the meal; that is, until the natural supplies furnish them with a preferable article. The average consumption of each colony is about two pounds of meal!

At the last annual Apiarian Convention in Germany, a cultivator recommended wheat flour as an excellent substitute for pollen. He says that in February, 1852, he used it with the best results. The bees *forsook the honey* which had been set out for them, and engaged actively in carrying in large quantities of the wheat flour, which was placed about twenty paces in front of the hives.

The construction of my hives, permits the flour to be placed, at once, where the bees can take it, without being compelled to waste their time in going out for it, or to suffer for the want of it, when the weather confines them at home.

The discovery of this substitute, removes a serious obstacle to the successful culture of bees. In many districts, there is a great abundance of honey for a few weeks in the season; and almost any number of colonies, which are strong when the honey harvest commences, will, in a good season, lay up sufficient stores for themselves, and a large surplus for their owners. In many of these districts, however, the supply of pollen is often so insufficient, that the new colonies of the previous year are found destitute of this article in the spring; and unless the season is early, and the weather unusually favorable, the production of brood is most seriously interfered with; thus the colony becomes strong too late to avail itself to the best advantage of the superabundant harvest of honey. (See remarks on the importance of having strong stocks early in the Spring.)

CHAPTER VII. ON THE ADVANTAGES WHICH OUGHT TO BE FOUND IN AN IMPROVED HIVE.

In this chapter, I shall enumerate certain very desirable, if not necessary, qualities of a good hive. I have neither the taste nor the time for the invidious work of disparaging other hives. I prefer inviting the attention of bee-keepers to the importance of these requisites; some of which, as I believe, are contained in no hive but my own. Let them be most carefully examined, and if they commend themselves to the enlightened judgment and good common sense of cultivators, let them be employed to test the comparative merits of the various kinds of hives in common use.

1. A good hive should give the Apiarian a perfect control over all the combs: so that any of them may be taken out at pleasure; and this, without cutting them, or enraging the bees.

This advantage is possessed by no hive in use, except my own; and it forms the very foundation of an improved and profitable system of bee-culture. Unless the combs are at the entire command of the Apiarian, he can have no effectual control over his bees. They swarm too much or too little, just as suits themselves, and their owner is almost entirely dependent upon their caprice.

2. It ought to afford suitable protection against extremes of heat and cold, sudden changes of temperature, and the injurious effects of dampness.

In winter, the interior of the hive should be dry, and not a particle of frost should ever find admission; and in summer, the bees should not be forced to work to disadvantage in a pent and almost suffocating heat. (See these points discussed in the Chapter on Protection.)

3. It should permit all necessary operations to be performed without hurting or killing a single bee.

Most hives are so constructed that it is impossible to manage them, without at times injuring or destroying some of the bees. The mere destruction of a few bees, would not, except on the score of humanity, be of much consequence, if it did not very materially increase the difficulty of managing them. Bees remember injuries done to any of their number, for some time, and generally find an opportunity to avenge them.

4. It should allow every thing to be done that is necessary in the most extensive management of bees, without incurring any serious risk of exciting their anger. (See Chapter on the Anger of Bees.)

5. Not a single unnecessary step or motion ought to be required of a single bee.

The honey harvest, in most locations, is of short continuance; and all the arrangements of the hive should facilitate, to the utmost, the work of the busy gatherers. Tall hives, therefore, and all such as compel them to travel with their heavy burdens through densely crowded combs, are very objectionable. The bees in my hive, instead of forcing their way through thick clusters, can easily pass into the surplus honey boxes, not only from any comb in the hive, but without traveling over the combs at all.

6. It should afford suitable facilities for inspecting, at all times, the condition of the bees.

When the sides of my hive are of glass, as soon as the outer cover is elevated, the Apiarian has a view of the interior, and can often at a glance, determine its condition. If the hive is of wood, or if he wishes to make a more thorough examination, in a few minutes every comb may be taken out, and separately inspected. In this way, the exact condition of every colony may always be easily ascertained, and nothing left, as in the common hives, to mere conjecture. This is an advantage, the importance of which it would be difficult to over estimate. (See Chapters on the loss of the queen, and on the Bee Moth.)

7. While the hive is of a size adapted to the natural instincts of the bee, it should be capable of being readily adjusted to the wants of small colonies.

If a small swarm is put into a large hive, they will be unable to concentrate their animal heat, so as to work to the best advantage, and will often become discouraged, and abandon their hive. If they are put into a small hive, its limited dimensions will not afford them suitable accommodations for increase. By means of my movable partition, my hive can, in a few moments, be adapted to the wants of any colony however small, and can, with equal facility, be enlarged from time to time, or at once restored to its full dimensions.

8. It should allow the combs to be removed without any jarring.

Bees manifest the utmost aversion to any sudden jar; for it is in this way, that their combs are loosened and detached. However firmly fastened the frames may be in my hive, they can all be loosened in a few moments, without injuring or exciting the bees.

9. It should allow every good piece of comb to be given to the bees, instead of being melted into wax. (See Chapter on Comb.)

10. The construction of the hive should induce the bees to build their combs with great regularity.

A hive which contains a large proportion of irregular comb, can never be expected to prosper. Such comb is only suitable for storing honey, or raising drones. This is one reason why so many colonies never flourish. A glance will often show that a hive contains so much drone comb, as to be unfit for the purposes of a stock hive.

11. It should furnish the means of procuring comb to be used as a guide to the bees, in building regular combs in empty hives; and to induce them more readily to take possession of the surplus honey receptacles.

It is well known that the presence of comb will induce bees to begin work much more readily than they otherwise Would: this is especially the case in glass vessels.

12. It should allow the removal of drone combs from the hive, to prevent the breeding of too many drones. (See remarks on Drones.)

13. It should enable the Apiarian, when the combs become too old, to remove them, and supply their place with new ones.

No hive can, in this respect, equal one in which, in a few moments, any comb can be removed, and the part which is too old, be cut off. The upper part of a comb, which is generally used for storing honey, will last without renewal for many years.

14. It ought to furnish the greatest possible security against the ravages of the Bee-Moth.

Neither before nor after it is occupied, ought there to be any cracks or crevices in the interior. All such places will be filled by the bees with propolis or bee-glue; a substance, which is always soft in the summer heat of the hive, and which forms a most congenial place of deposit for the eggs of the moth. If the sides of the hive are of glass, and the corners are run with a melted mixture, three parts rosin, and one part bees-wax, the bees will waste but little time in gathering propolis, and the bee-moth will find but little chance for laying her eggs, even if she should succeed in entering the hive.

My hives are so constructed, that if made of wood, they may be thoroughly painted inside and outside, without being so smooth as to annoy the bees; for they travel over the frames to which the combs are attached; and thus whether the inside surface is glass or wood, it is not liable to crack, or warp, or absorb moisture, after the hive is occupied by the bees. If the hives are painted inside, it should be done sometime before they are used. If the interior of the wooden hive is brushed with a very hot mixture of the rosin and bees-wax, the hives may be used immediately.

15. It should furnish some place accessible to the Apiarian, where the bee-moth can be tempted to deposit her eggs, and the worms, when full grown, to wind themselves in their cocoons. (See remarks on the Bee-Moth.)

16. It should enable the Apiarian, if the bee-moth ever gains the upper hand of the bees, to remove the combs, and expel the worms. (See Bee-Moth.)

17. The bottom board should be permanently attached to the hive; for if this is not done, it will be inconvenient to move the hive when bees are in it, and next to impossible to prevent the depredations of moths and worms.

Sooner or later, there will be crevices between the bottom board and the sides of the hive, through which the moths will gain admission, and under which the worms, when fully grown, will retreat to spin their webs, and to be changed into moths, to enter in their turn, and lay their eggs. Movable bottom hoards are a great nuisance in the Apiary, and the construction of my hive, which enables me entirely to dispense with them, will furnish a very great protection against the bee-moth. There is no place where they can get in, except at the entrance for the bees, and this may be contracted or enlarged, to suit the strength of the colony; and from its peculiar shape, the bees are enabled to defend it against intruders, with the greatest advantage.

18. The bottom-board should slant towards the entrance, to assist the bees in carrying out the dead, and other useless substances; to aid them in defending themselves against robbers; to carry off all moisture; and to prevent the rain and snow from beating into the hive. As a farther precaution against this last evil, the entrance ought to be under a covered way, which should not, at once lead into the interior.

19. The bottom-board should be so constructed that it may be readily cleared of dead bees in cold weather, when the bees are unable to attend to this business themselves.

If suffered to remain, they often become mouldy, and injure the health of the colony. If the bees drag them out, as they will do, if the weather moderates, they often fall with them on the snow, and are so chilled that they never rise again; for a bee generally retains its hold in flying away with the dead, until both fall to the ground.

20. No part of the interior of the hive should be below the level of the place of exit.

If this principle is violated, the bees must, at great disadvantage, drag their dead, and all the refuse of the hive, *up hill*. Such hives will often have their bottom boards covered with small pieces of comb, bee-bread, and other impurities, in which the moth delights to lay her eggs; and which furnished her progeny with a most congenial nourishment, until they are able to get access to the combs.

21. It should afford facilities for feeding the bees both in warm and cold weather.

In this respect, my hive has very unusual advantages. Sixty colonies in warm weather may, in an hour, be fed a quart each, and yet no feeder be used, and no risk incurred from robbing bees. (See Chapter on Feeding.)

22. It should allow of the easy hiving of a swarm, without injuring any of the bees, or risking the destruction of the queen. (See Chapter on Natural Swarming, and Hiving.)

23. It should admit of the safe transportation of the bees to any distance whatever.

The permanent bottom-board, the firm attachment of the combs, each to a separate frame, and the facility with which, in my hive, any amount of air can be given to the bees when shut up, most admirably adapt it to this purpose.

24. It should furnish the bees with air when the entrance is shut; and the ventilation for this purpose ought to be unobstructed, even if the hives should be buried in two or three feet of snow. (See Chapter on Protection.)

25. A good hive should furnish facilities for enlarging, contracting, and closing the entrance; so as to protect the bees against robbers, and the bee-moth; and when the entrance is altered, the bees ought not to lose valuable time in searching for it, as they must do in most hives. (See Chapters on Ventilation, and on Robbing.)

26. It should give the bees the means of ventilating their hives,

without enlarging the entrance too much, so as to expose them to moths and robbers, and to the risk of losing their brood by a chill in sudden changes of weather. (See Chapter on Ventilation.)

To secure this end, the ventilators must not only be independent of the entrance, but they must owe their efficiency mainly to the co-operation of the bees themselves, who thus have a free admission of air only when they want it. To depend on the opening and shutting of the ventilators by the bee-keeper, is entirely out of the question.

27. It should furnish facilities for admitting at once, a large body of air; so that in winter, or early spring, when the weather is at any time unusually mild, the bees may be tempted to fly out and discharge their fæces. (See Chapter on Protection.)

If such a free admission of air cannot be given to hives which are thoroughly protected against the cold, the bees may lose a favorable opportunity of emptying themselves; and thus be more exposed than they otherwise would, to suffer from diseases resulting from too long confinement. A very free admission of air is also desirable when the weather is exceedingly hot.

28. It should enable the Apiarian to remove the excess of bee-bread from old stocks.

This article always accumulates in old hives, so that in the course of time, many of the combs are filled with it, thus unfitting them for the rearing of brood, and the reception of honey. Young stocks, on the other hand, will often be so deficient in this important article, that in the early part of the season, breeding will be seriously interfered with. By means of my movable frames, the excess of old colonies may be made to supply the deficiency of young ones, to the mutual benefit of both. (See Chapter on Pollen.)

29. It should enable the Apiarian, when he has removed the combs from a common hive, to place them with the bees, brood, honey and bee-bread, in the improved hive, so that the bees may be able to attach them in their natural positions. (See directions for transferring bees from an old hive.)

30. It should allow of the easy and safe dislodgement of the bees from the hive.

This requisite is especially important to secure the union of colonies, when it becomes necessary to break up some of the stocks. (See remarks on the Union of Stocks.)

31. It should allow the heat and odor of the main hive, as well as the bees themselves, to pass in the freest manner, to the surplus honey receptacles.

In this respect, all the hives with which I am acquainted, are more or less deficient: the bees are forced to work in receptacles difficult of access, and in which, especially in cool nights, they find it impossible to keep up the animal heat necessary for comb-building. Bees cannot, in such hives, work to advantage in glass tumblers, or other small vessels. One of the most important arrangements of my hive, is that by which the heat ascends into all the receptacles for storing honey, as naturally and almost as easily as the warmest air ascends to the top of a heated room.

32. It should permit the surplus honey to be taken away, in the most convenient, beautiful and salable forms, at any time, and without any risk of annoyance from the bees.

In my hives, it may be taken in tumblers, glass boxes, wooden boxes small or large, earthen jars, flower-pots; in short, in any kind of receptacle which may suit the fancy, or the convenience of the bee-keeper. Or all these may be dispensed with, and the honey may be taken from the interior of the main hive, by removing the frames with loaded combs, and supplying their place with empty ones.

33. It should admit of the easy removal of all the good honey from the main hive, that its place may be supplied with an inferior article.

Bee-Keepers who have but few colonies, and who wish to secure the largest yield, may remove the loaded combs from my hive, slice off the covers of the cells, drain out the honey, and restore the empty combs, into which, if the season of gathering is over, they can first pour the cheap foreign honey for the use of the bees.

34. It should allow, when quantity not quality is the object, the largest amount of honey to be gathered; so that the surplus of strong colonies may, in the Fall, be given to those which have not a sufficient supply.

By surmounting my hive with a box of the same dimensions, the combs may all be transferred to this box, and the bees, when they commence building, will descend and fill the lower frames, gradually using the upper box, as the brood is hatched out, for storing honey. In this way, the largest possible yield of honey may be secured, as the bees always prefer to continue their work below, rather than above the main hive, and will never swarm, when allowed in season, ample room in this direction. The combs in the upper box, containing a large amount of bee-bread and being of a size adapted to the breeding of workers, will be all the better for aiding weak colonies.

35. It should compel, when desired, the force of the colony to be mainly directed to raising young bees; so that brood may be on hand to form new colonies, and strengthen feeble stocks. (See Chapter on Artificial Swarming.)

36. It ought, while well protected from the weather, to be so constructed, that in warm, sunny days in early spring, the influence of the sun may be allowed to penetrate and warm up the hive, so as to encourage early breeding. (See Chapter on Protection.)

37. The hive should be equally well adapted to be used as a swarmer, or non-swarmer.

In my hives bees may be allowed, if their owner chooses, to swarm just as they do in common hives, and be managed in the usual way. Even on this plan, the great protection against the weather which it affords, and the command over all the combs, will be found to afford great advantages. (See Natural Swarming.)

Non-swarming hives managed in the ordinary way are liable, in spite of all precautions, to swarm very unexpectedly, and if not closely watched, the swarm is lost, and with it the profit of that season. By having the command of the combs, the queen in my hives can always be caught and deprived of her wings; thus she cannot go off with a swarm, and they will not leave without her.

38. It should enable the Apiarian, if he allows his bees to swarm, and wishes to secure surplus honey, to prevent them from throwing more than one swarm in a season.

Second and third swarms must be returned to the old stock, if the largest quantities of surplus honey are to be realized. It is troublesome to watch them, deprive them of their queens, and restore them to the parent hive. They often issue with new queens again and again; and waste, in this way, both their own time, and that of their keeper. "An ounce of prevention is worth a pound of cure." In my hives, as soon as the first swarm has issued, and been hived, all the queen cells except one, in the hive from which it came, may be cut out, and thus all after-swarming will very easily and effectually be prevented. (See Chapter on Artificial Swarming, for the use to which these supernumerary queens may be put.) When the old stock is left with but one queen, she runs no risk of being killed or crippled in a contest with rivals. By such contests, a colony is often left without a queen, or in possession of one which is too much maimed to be of any service. (See Chapter on the Loss of the Queen.)

39. A good hive should enable the Apiarian, if he relies on natural swarming, and wishes to multiply his colonies as fast as possible, to make vigorous stocks of all his small after-swarms.

Such swarms contain a young queen, and if they can be judiciously strengthened, usually make the best stock hives. If hived in a common hive, and left to themselves, unless very early, or in very favorable seasons, they seldom thrive. They generally desert their hives, or perish in the winter. If they are small, they cannot be made powerful, even by the most generous feeding. There are too few bees to build comb, and take care of the eggs which a healthy queen can lay; and when fed, they are apt to fill with honey, the cells in which young bees ought to be raised; thus making the kindness of their owner serve only to hasten their destruction. My hives enable me to supply all such swarms at once with combs containing bee-bread, honey and brood almost mature. They are thus made strong, and flourish as well, nay, often better than the first swarms which have an old queen, whose fertility is generally not so great as that of a young one.

40. It should enable the Apiarian to multiply his colonies with a certainty and rapidity which are entirely out of the question, if he depends upon natural swarming. (See Chapter on Artificial Swarming.)

41. It should enable the Apiarian to supply destitute colonies with the means of obtaining a new queen.

Every Apiarian would find it, for this reason, if for no other, to his advantage to possess, at least, one such hive. (See Chapters on Physiology, and loss of Queen.)

42. It should enable him to catch the queen, for any purpose; especially to remove an old one whose fertility is impaired by age, that her place may be supplied with a young one. (See Chapter on Artificial Swarming.)

43. While a good hive is adapted to the wants of those who desire to enter upon bee-keeping on a large scale, or at least to manage their colonies on the most improved plans, it ought to be suited to the wants of those who are too timid, too ignorant, or for any reason indisposed, to manage them in any other than the common way.

44. It should enable a single individual to superintend the colonies

of many different persons.

Many would like to keep bees, if they could have them taken care of, by those who would undertake their management, just as a gardener does the gardens and grounds of his employers. No person can agree to do this with the common hives. If the bees are allowed to swarm, he may be called in a dozen different directions, and if any accident, such as the loss of a queen, happens to the colonies of his customers, he can apply no remedy. If the bees are in non-swarming hives, he cannot multiply the stocks when this is desired.

On my plan, gentlemen who desire it, may have the pleasure of witnessing the industry and sagacity of this wonderful insect, and of gratifying their palates with its delicious stores, harvested on their own premises, without incurring either trouble or risk of injury.

45. All the joints of the hive should be water-tight, and there should be no doors or slides which are liable to shrink, swell, or get out of order.

The importance of this will be sufficiently obvious to any one who has had the ordinary share of vexatious experience in the use of such fixtures.

46. It should enable the bee-keeper entirely to dispense with sheds, and costly Apiaries; as each hive when properly placed, should alike defy, heat or cold, rain or snow. (See Chapter on Protection.)

47. It should allow the contents of a hive, bees, combs and all, to be taken out; so that any necessary repairs may be made.

This may be done, with my hives, in a few minutes. "A stitch in time saves nine." Hives which can be thoroughly overhauled and repaired, from time to time, if properly attended to, will last for generations.

48. The hive and fixtures should present a neat and attractive appearance, and should admit, when desired, of being made highly ornamental.

49. The hives ought not to be liable to be blown down in high winds.

My hives, being very low in proportion to their other dimensions, it would require almost a hurricane to upset them.

50. It should enable an Apiarian who lives in the neighborhood of human pilferers, to lock up the precious contents of his hives, in some cheap, simple and convenient way.

A couple of padlocks with some cheap fixtures, will suffice to secure a long range of hives.

51. A good hive should be protected against the destructive ravages of mice in winter.

It seems almost incredible that so puny an animal should dare to invade a hive of bees; and yet not infrequently they slip in when the bees are compelled by the cold to retreat from the entrance. Having once found admission, they build themselves a nest in their comfortable abode, eat up the honey, and such bees as are too much chilled to make any resistance; and fill the premises with such an abominable stench, that on the approach of warm weather, the bees often in a body abandon their desecrated home. As soon as the cold weather approaches, all my hives may have their entrances either entirely closed, or so contracted that a mouse cannot gain admission.

52. A good hive should have its alighting board constructed so as to shelter the bees against wind and wet, and thus to facilitate to the utmost their entrance when they come home with their heavy burdens.

If this precaution is neglected, much valuable time and many lives will be sacrificed, as the colony cannot be encouraged to use to the best advantage the unpromising days which so often occur in the working season.

I have succeeded in arranging my alighting board in such a manner that the bees are sheltered against wind and wet, and are able to enter the hive with the least possible loss of time.

53. A well constructed hive ought to admit of being shut up in winter, so as to consign the bees to darkness and repose.

Nothing can be more hazardous than to shut up closely an ill protected hive. Even if the bees have an abundance of air, it will not answer to prevent them from flying out, if they are so disposed. As soon as the warmth penetrating their thin hives tempts them to fly, they crowd to the entrance, and if it is shut, multitudes worry themselves to death in trying to get out, and the whole colony is liable to become diseased.

In my hives as soon as the bees are shut up for Winter, they are most effectually protected against all atmospheric changes, and never *desire* to leave their hives until the entrances are again opened, on the return of suitable weather. Thus they pass the Winter in a state of almost absolute repose; they eat much less honey[12] than when wintered on the ordinary plan; a much smaller number die in the hives; none are lost upon the snow, and they are more healthy, and commence breeding much earlier than they do in the common hives. As some of the holes into the Protector are left open in Winter, any bee that is diseased and wishes to leave the hive can do so. Bees when diseased have a strange propensity to leave their hives, just as animals when sick seek to retreat from their companions; and in Summer such bees may often be seen forsaking their home to perish on the ground. If all egress from the hive in Winter is prevented, the diseased bees will not be able to comply with an instinct which urges them "To leave their country for their country's good."

54. It should possess all these requisites without being too costly for common bee-keepers, or too complicated to be constructed by any one who can handle simple tools: and they should be so combined that the result is a simple hive, which any one can manage who has ordinary intelligence on the subject of bees.

I suppose that the very natural conclusion from reading this long list of desirables, would be that no single hive can combine them all, without being exceedingly complicated and expensive. On the contrary, the simplicity and cheapness with which my hive secures all these results, is one of its most striking peculiarities, the attainment of which has cost me more study than all the other points besides. As far as the bees are concerned, they can work in this hive with even greater facility than in the simple old-fashioned box, as the frames are left rough by the saw, and thus give an admirable support to the bees when building their combs; and they can enter the spare honey boxes, with even more ease than if they were merely continuations of the main hive.

There are a few desirables to which my hive makes not the slightest pretensions! It promises no splendid results to those who purchase it, and yet are too ignorant, or too careless to be entrusted with the management of bees. In bee-keeping, as in other things, a man must first understand his business, and then proceed on the good old maxim, that "the hand of the diligent maketh rich."

It possesses no talismanic influence by which it can convert a bad situation for honey, into a good one; or give the Apiarian an abundant harvest whether the season is productive or otherwise.

It cannot enable the cultivator rapidly to multiply his stocks, and yet to secure, the same season, surplus honey from his bees. As well might the breeder of poultry pretend that he can, in the same year, both raise the greatest number of chickens, and sell the largest number of eggs.

Worse than all, it cannot furnish the many advantages enumerated, and yet be made in as little time, or quite as cheap as a hive which proves, in the end, to be a very dear bargain.

I have not constructed my hive in accordance with crude theories, or mere conjectures, and then insisted that the bees must flourish in such a fanciful contrivance; but I have studied, for many years, most carefully, the nature of the honey-bee; and have diligently compared my observations with those of writers and practical cultivators, who have spent their lives in extending the sphere of Apiarian knowledge; and as the result, have endeavored to adapt my hive to the actual wants and habits of the bee; and to remedy the many difficulties with which I have found its successful culture to be beset. And more than this, I have actually tested by experiments long continued and on a large scale, the merits of this hive, that I might not deceive both myself and others, and add another to the many useless contrivances which have deluded and disgusted a credulous public. I would, however, most earnestly repudiate all claims to having devised a "perfect bee-hive." Perfection can belong only to the works of the great Creator, to whose Omniscient eye, all causes and effects with all their relations, were present, when he spake, and from nothing formed the universe and all its glorious wonders. For man to stamp upon any of his own works, the label of perfection, is to show both his folly and presumption.

It must be confessed that the culture of bees is at a very low ebb in our country, when thousands can be induced to purchase hives which are in most glaring opposition not only to the true principles of Apiarian knowledge, but often, to the plainest dictates of simple common sense. Such have been the losses and disappointments of deluded purchasers, that it is no wonder that they turn from everything offered in the shape of a patent bee-hive, as a miserable humbug, if not a most bare-faced cheat.

I do not hesitate to say that those old-fashioned bee-keepers, who have most steadily refused to meddle with any novelties, and who have used hives of the very simplest construction, or at least such as are only one remove from the old straw hive, or wooden box, have, as a general thing, realized by far the largest profits in the management of bees. They have lost neither time, money nor bees, in the vain hope of obtaining any unusual results from hives, which, in the very nature of the case, can secure nothing really in advance of what can be accomplished by a simple box-hive with an upper chamber.

A hive of the simplest possible construction, is only a close imitation of the abode of bees in a state of nature; being a mere hollow receptacle in which they are protected from the weather, and where they can lay up their stores.

An improved hive is one which contains, in addition, a separate apartment in which the bees can be induced to lay up the surplus portion of their stores, for the use of their owner. All the various hives in common use, are only modifications of this latter hive, and, as a general rule, they are bad, exactly in proportion as they depart from it. Not one of them offers any remedy for the loss of the queen, or indeed for most of the casualties to which bees are exposed: they form no reliable basis for any new system of management; and hence the cultivation of bees, is substantially where it was, fifty years ago, and the Apiarian as entirely dependent as ever, upon all the whims and caprices of an insect which may be made completely subject to his control.

No hive which does not furnish a thorough control over every comb, can be considered as any substantial advance on the simple improved or chamber hive. Of all such hives, the one which with the least expense, gives the greatest amount of protection, and the readiest access to the spare honey boxes, is the best.

Having thus enumerated the tests to which all hives ought to be subjected, and by which they should stand or fall, I submit them to the candid examination of practical, common sense bee-keepers, who have had the largest experience in the management of bees, and are most conversant with the evils of the present system; and who are therefore best fitted to apply them to an invention, which, if I may be pardoned for using the enthusiastic language of an experienced Apiarian on examining its practical workings, "introduces, not simply an *improvement*, but a *revolution* in bee-keeping."

CHAPTER VIII. PROTECTION AGAINST EXTREMES OF HEAT AND COLD, SUDDEN AND SEVERE CHANGES OF TEMPERATURE, AND DAMPNESS IN THE HIVES.

I specially invite a careful perusal of this chapter, as the subject, though of the very first importance in the management of bees, is one to which but little attention has been given by the majority of cultivators.

In our climate of great and sudden extremes, many colonies are annually injured or destroyed by undue exposure to heat or cold. In Summer, thin hives are often exposed to the direct heat of the sun, so that the combs melt, and the bees are drowned in their own sweets. Even if they escape utter ruin, they cannot work to advantage in the almost suffocating heat of their hives.

But in those places where the Winters are both long and severe, it is much more difficult to protect the bees from the cold than from the heat. Bees are not, as some suppose, in a *dormant*, or *torpid* condition in Winter. It must be remembered that they were intended to live in colonies, in Winter, as well as Summer. The wasp, hornet, and other insects which do not live in families in the Winter, lay up no stores for cold weather, and are so organized as to be able to endure in a torpid state, a very low temperature; so low that it would be certain death to a honey-bee, which when frozen, is as surely killed as a frozen man.

As soon as the temperature of the hives falls too low for their comfort, the bees gather themselves into a more compact body, to preserve to the utmost, their animal heat; and if the cold becomes so great that this will not suffice, they keep up an incessant, tremulous motion, accompanied by a loud humming noise; in other words, they take active exercise in order to keep warm! If a thermometer is pushed up among them, it will indicate a high temperature, even when the external atmosphere is many degrees below zero. When the bees are unable to maintain the necessary amount of animal heat, an occurrence which is very common with small colonies in badly protected hives, then, as a matter of course, they must perish.

Extreme cold, when of long continuance, very frequently destroys colonies in thin hives, even when they are strong both in bees and honey. The inside of such hives, is often filled with frost, and the bees, after eating all the food in the combs in which they are clustered, are unable to enter the frosty combs, and thus starve in the midst of plenty. The unskilfull bee-keeper who finds an abundance of honey in the hives, cannot conjecture the cause of their death.

If the cold merely destroyed feeble colonies, or strong ones only now and then, it would not be so formidable an enemy; but every year, it causes many of the most flourishing stocks to perish by starvation. The extra quantity of food which they are compelled to eat, in order to keep up their heat in their miserable hives, is often the turning point with them, between life and death. They starve, when with proper protection, they would have had food enough and to spare.

But some one may say, "What possible difference can the kind of hives in which bees are kept make in the quantity of food which they will consume?" Enough, I would reply, in some single winters, to pay the difference between a good hive and a bad one!

I cannot move my finger, or wink my eye-lids without some waste of muscle, however small; for it is a well-ascertained law in our animal economy, that all *muscular exertion* is attended with a corresponding *waste* of muscular fibre. Now this waste must be supplied by the consumption of food, and it would be as unreasonable to expect constant heat from a stove without fresh supplies of fuel, as incessant muscular activity from an insect, without a supply of food proportioned to that activity. If then we can contrive any way to keep our bees in almost perfect quiet during the Winter, we may be certain that they will need much less food than when they are constantly excited.

In the cold Winter of 1851-2, I kept two swarms in a perfectly dry and dark cellar, where the temperature was remarkably uniform, seldom varying two degrees from 50° of Fahrenheit; and I found that the bees ate very little honey. The hives were of glass, and the bees, when examined from time to time, were found clustered in almost death-like repose. If these bees had been exposed in thin hives in the open air, they would, in all probability, have eaten four times as much; for whenever the sun shone upon them, or the atmosphere was unusually warm, they would have been roused to injurious activity, and the same would have been the case, when the cold was severe. Exposed to sudden changes and severe cold, they would have been in almost perpetual motion, and must have been compelled to consume a largely increased quantity of food. In this way, many colonies are annually starved to death, which if they had been better protected, would have survived to gladden their owner with an abundant harvest. This protection, as a general thing, must be given to them in the open air, for it is a very rare thing, to meet with a cellar which is dry enough to prevent the combs from moulding, and the bees from becoming diseased.

Bees never, unless diseased, discharge their fæces in the hive; and the want of suitable protection, by exciting undue activity, and compelling them to eat more freely, causes their bodies to be greatly distended with accumulated fæces. On the return of warm weather, bees in this condition being often too feeble to fly, crawl from their hives, and miserably perish.

I must notice another exceedingly injurious effect of insufficient protection, in causing the *moisture* to settle upon the cold top and sides of the interior of the hive, from whence it drips upon the bees. In this way, many of their number are chilled and destroyed, and often the whole colony is infected with dysentery. Not infrequently, large portions of the comb are covered with mould, and the whole hive is rendered very offensive.

This dampness which causes what may be called a *rot* among the bees, is one of the worst enemies with which the Apiarian in a cold climate, has to contend, as it weakens or destroys many of his best colonies. No extreme of cold ever experienced in latitudes where bees flourish, can destroy a strong colony well supplied with honey, except indirectly, by confining them to empty combs. They will survive our coldest winters, in thin hives raised on blocks to give a freer admission of air, or even in suspended hives, without any bottom-board at all. Indeed, in cold weather, a *very free* admission of air is necessary in such hives, to prevent the otherwise ruinous effects of frozen moisture; and hence the common remark that bees require as much or more air in Winter than in Summer.

When bees, in unsuitable hives, are exposed to all the variations of the external atmosphere, they are frequently tempted to fly abroad if the weather becomes unseasonably warm, and multitudes are lost on the *snow*, at a season when no young are bred to replenish their number, and when the loss is most injurious to the colony.

From these remarks, it will be obvious to the intelligent cultivator, that protection against extremes of heat and cold, is a point of the VERY FIRST IMPORTANCE; and yet this is the very point, which, in proportion to its importance, has been most overlooked. We have discarded, and very wisely, the straw hives of our ancestors; but such hives, with all their faults, were comparatively warm in Winter, and cool in Summer. We have undertaken to keep bees, where the cold of Winter, and the heat of Summer are alike intense; and where sudden and severe changes are often fatal to the brood: and yet we blindly persist in expecting success under circumstances in which any marked success is well nigh impossible.

That our country is eminently favorable to the production of honey, cannot be doubted. Many of our forests abound With colonies which are not only able to protect themselves against all their enemies, the dreaded bee-moth not excepted, but which often amass prodigious quantities of honey. Nor are such colonies found merely in *new* countries. They exist frequently in the very neighborhood of cultivators whose hives are weak and impoverished, and who impute to a decay of the honey resources of the country, the inevitable consequences of their own irrational system of management. It will not be without profit, to consider briefly under what circumstances these wild colonies flourish, and how they are protected against sudden and extreme changes of temperature.

Snugly housed in the hollow of a tree whose thickness and decayed interior are such admirable materials for excluding atmospheric changes, the bees in Winter are in a state of almost absolute repose. The entrance to their abode is generally very small in proportion to the space within; and let the weather out of doors vary as it may, the inside temperature is very uniform. These natural hives are dry, because the moisture finds no cold or icy top, or sides, on which to condense, and from which it must drip upon the bees, destroying their lives, or enfeebling their health, by filling the interior of their dwelling with mould and dampness. As they are very quiet, they eat but little, and hence their bodies are not distended and diseased by accumulated fæces. Often they do not stir from their hollows, from November until March or April; and yet they come forth in the Spring, strong in numbers, and vigorous in health. If at any time in the winter season, the warmth is so great as to penetrate their comfortable abodes, and to tempt them to fly, when they venture out, they find a balmy atmosphere in which they may disport with impunity. In the Summer, they are protected from the heat, not merely by the thickness of the hollow tree, but by the leafy shade of overarching branches, and the refreshing coolness of a forest home.

The Russian and Polish bee-keepers, living in a climate whose winters are much more severe than our own, are among the largest and most successful cultivators of bees, many of them numbering their colonies by hundreds, and some even by thousands!

They have, with great practical sagacity, imitated as closely as possible, the conditions under which bees are found to flourish so admirably in a state of nature. We are informed by Mr. Dohiogost, a Polish writer, that his countrymen make their hives of the best plank, and never less than an inch and a half in thickness. The shape is that of an old-fashioned churn, and the hive is covered on the outside, halfway down, with twisted rope cordage, to give it greater protection against extremes of heat and cold. The hives are placed in a dry situation, directly upon the hard earth, which is first covered with an inch or two of clean, dry sand. Chips are then heaped up all around them, and covered with earth banked up in a sloping direction to carry off the rain. The entrance is at some distance above the bottom, and is a triangle, whose sides are only one inch long. In the winter season, this entrance is contracted so that only one bee can pass at a time. Such a hive, with us, as it does not furnish the honey in convenient, beautiful and salable forms, would not meet the demands of our cultivators. Still, there are some very important lessons to be learned from it, by all who keep bees in regions of cold winters, and hot summers. It shows the importance which some of the largest Apiarians in the world, attach to protection; practical, common sense men, whose heads have not been turned, as some would express it, by modern theories and fanciful inventions. They cultivate their bees almost in a state of nature, and their experience on what we would term a gigantic scale, ought to convince even the most incredulous, of the folly of pretending to keep bees, in the miserably thin and unprotected hives to which we have been accustomed.

But how, it will be asked, can bees live in Winter, in a hive so closely shut up as the Polish hive? They do live in such hives, and prosper, just as they do in hollow trees, with only one small entrance. It is well known that bees have flourished when their hives were buried in Winter, and under circumstances in which but a very small amount of air could possibly gain admission to them. Bees, when kept in a *dry* place, in properly protected hives and in a state of almost perfect repose, need only a small supply of air; and the objection that those cultivators among us, who shut up their colonies very closely in Winter, are almost sure to lose them, is of no weight; because the majority of our hives are so deficient in protection, that if they are too closely shut up, "the breath of the bees," condensing and freezing upon the inside, and afterwards thawing, causes the combs to mould, and the bees to become diseased; just as many substances mould and perish when kept in a close, damp cellar.

We are now prepared to discuss the question of protection in its relations to the construction of hives. We have seen how it is furnished to the bees in the Polish hives, and in the decayed hollows of trees. If the Apiarian chooses, he can imitate this plan by constructing his hives of very thick plank: but such hives would be clumsy, and with us, expensive. Or he may much more effectually reach the same end, by making his hives double, so as to enclose an air space all around, which in Winter may be filled with charcoal, plaster of Paris, straw, or any good non-conductor, to enable the bees to preserve with the least waste, their animal heat. I prefer to pack the air-space with plaster of Paris, as it is one of the very best non-conductors of heat, being used in the manufacture of the celebrated Salamander fire-proof safes. Hives may be constructed in this way, which without great expense, may be much better protected than if they were made of six-inch plank. As the price of glass is very low, I prefer to construct the inside of my doubled hives of this material. When a number of hives are to be made, as the lowest price glass will answer every purpose, I can furnish a given amount of protection cheaper with glass than wood, while the glass possesses some most decided advantages over any other material. The hives are lighter and more compact, than when made of doubled wood, and can be more easily moved, while the Apiarian can gratify his rational curiosity, and inspect at all times, the condition of his stocks. The very interest inspired by being able to see what they are doing, will go far to protect them from that indifference and neglect, which is so often fatal to their prosperity. The way in which I make my hives, not only protects the bees against extremes of heat and cold, but it guards them very effectually, against the injurious and often fatal effects of condensed moisture. By means of my movable frames, the combs are prevented from being attached to the sides, top or bottom of the hive; they are in fact, suspended in the air. If now the dampness can be prevented from condensing any where, *over* the bees, so that it may not drip upon their combs, and if it can be easily discharged from the hive wherever it may collect, it cannot, under any circumstances, seriously annoy them. Such are the arrangements in my hives, that but very little moisture forms in them, and all that does, is deposited on the sides in preference to any other part of the interior; just as it is upon the colder walls or windows, rather than the ceiling of a room. But as the combs are kept away from the sides, this moisture cannot annoy the bees; nor can it penetrate the glass as it does unpainted wood or straw, thus causing a more protracted dampness; it must run down their smooth surfaces, and fall upon the bottom-board, from whence it can be easily discharged from the hive. By packing in winter, the necessary amount of protection is secured for the top and sides of the hive, and the very worst property of glass, (its parting so rapidly with heat,) is changed into one of the very best for the purposes of a bee-hive. I prefer not only to make the sides of my hive of glass, but of *double* glass, with an air space of about an inch between the two panes of glass. The extra cost[13] of this construction will be amply repaid by the additional protection given to the bees. It will be absolutely impossible for any frost ever to penetrate through this air space, and the packing between the outside case and the main hive. The combs in such a hive cannot be melted down, even if the hive is exposed to the reflected and concentrated heat of a blazing sun: the same construction which secures them against the cold of Winter, equally protecting them from the heat of Summer. There is one disadvantage to which all well protected hives of the ordinary construction, are exposed. In the Spring of the year, it is exceedingly desirable that the warmth of the sun should penetrate the hives, to encourage the bees in early breeding; but the very arrangement which protects them from cold, often interferes with this. A bee-hive is thus like a cellar, warm in Winter, and cool in Summer; but often unpleasantly cool in the early Spring, when the atmosphere out of doors is warm and delightful. In my hive, this difficulty is easily remedied. In the Spring, as soon as the bees begin to fly, on warm, sun-shiny days, the upper part of the outside case is removed, so that the genial heat of the sun can penetrate to every part of the hive. The cover must be replaced while the sun is still shining, so that the hives may be shut up while they are warm. The labor of doing this, need occupy only a few minutes daily, and as soon as warm weather fairly sets in, it may be dispensed with. It may be performed without any risk, by a woman or a boy.

If the hive is of glass, it will warm up all the better, and as the combs are on frames, they cannot be melted or injured by the heat. It is a serious objection to most covered Apiaries, that they do not permit the hives to receive the genial heat of the sun at a period of the year when instead of injuring the bees, it exerts a most powerful influence in developing their brood.

This is one among many reasons why I have discarded them, and why I prefer to construct my hives in such a manner that they need no extra covering, but stand exposed to the full influence of the sun. I have known strong colonies which have survived the Winter in thin hives, to increase rapidly and swarm early, because of the stimulating effect of the sun; while others, deprived of this influence, in dark bee houses and well protected hives, have sometimes disappointed the hopes of their owners. Although my glass hives are very beautiful, and most admirably protected, still hives of doubled wood may often be built to better advantage by those who construct their own hives, and they can be made to furnish any desirable amount of protection.

Enclosed Apiaries are at best but nuisances: they soon become lurking-places for spiders and moths; and after all the expense wasted on their construction, afford, but little protection against extreme cold.

I have been thus particular on the subject of protection, in order to convince every bee keeper who exercises common sense, that thin hives ought to be given up, if either pleasure or profit is sought from his bees. Such hives an enlightened Apiarian could not be persuaded to purchase, and he would consider them too expensive in their waste of honey and bees, to be worth accepting, even as a gift. Many strong colonies which are lodged in badly protected hives, often consume in extra food, in a single hard winter, more than enough to pay the difference between the first cost of a good hive over a bad one. In the severe winter of 1851-2, many cultivators lost nearly all their stocks, and a large part of those which survived, were too much weakened to be able to swarm. And yet these same miserable hives, after accomplishing the work of destruction on one generation of bees, are reserved to perform the same office for another. And this some call economy!

I am well aware of the question which many of my readers have for some time been ready to ask of me. Can you make one of your well protected hives as cheaply as we construct our common hives? I would remind such questioners, that it is hardly possible to build a well protected house as cheaply as a barn.

And yet by building my hives in solid structures, three together,

I am able to make them for a very moderate price, and still to give them even better protection than when they are constructed singly. If they are not built of doubled materials they can be made for as little money as any other patent hive, and yet afford much greater protection; as the combs touch neither the top, bottom nor sides of the hive. I recommend however a construction, which although somewhat more costly at first, is yet much cheaper in the end.

Such is the passion of the American people for cheapness in the first cost of an article, even at the evident expense of dearness in the end, that many, I doubt not, will continue to lodge their bees in thin hives, in spite of their conviction of the folly of so doing; just as many of our shrewdest Yankees build thin wooden houses, in the cold climate of New England, or plaster their stone or brick ones directly on the wall, when the extra cost of fuel to warm them, far exceeds the interest on the additional expense which would be necessary to give them the requisite protection; to say nothing of the doctors' bills, and fatal diseases which can be traced often to the dreary barns or damp vaults which they build, and call houses!

PROTECTOR.

I attach very great importance to the way in which I give the bees effectual protection against extremes of heat and cold, and sudden changes of temperature, without removing them from their stands, or incurring the expense and disadvantages of a covered Bee-House. This I accomplish by means of what I shall call a *Protector* which is constructed substantially as follows.

Select a dry and suitable location for the bees, where they will not be disturbed, or prove an annoyance to others. If possible, let it be in full sight of the sitting room, so that they may be seen in case of swarming; and let it face the South-East, and be well protected from the force of strong winds. Dig a trench, about two feet deep; its length should depend upon the number of hives to be accommodated; and its breadth should be such that when it is properly walled up, it should measure from the outside top of one wall to another, just sufficient to receive the bottom of the hive. The walls, may be built of refuse brick or stones, and should be about four feet high from the foundation; the upper six inches being built of good brick, and the back wall about two inches higher than the front one, so as to give the bottom-board of the hives, the proper slant towards the entrance. At one end of this Protector, a wooden chimney should be built, and if the number of hives is great, there should be one at each end, admitting air in Winter, and yet excluding rain and snow. The earth which is thrown out in digging, should be banked up against the walls as high as the good brick, and in a slope which, when grassed over, may be easily mowed with a common scythe. The slope on the back should be more perpendicular than in front so as not to be in the way when operating upon the hives.

The bottom may be covered with an inch or two of clean sand and in winter with straw. In Summer, the ends are left open, so that a free current of air may pass through, while in Winter, they are properly banked up; and straw, evergreen boughs, or any other material, suitable for excluding frost, may if necessary, be placed all around the outside of the Protector. Such an arrangement will be found very cheap, when compared with a Bee-House or covered Apiary, and may be made both neat and highly ornamental. It may be constructed of wood by those who desire something still cheaper, and any one who can handle a spade, hammer, plane and saw, can make for himself a structure on which a hundred hives may stand, at less expense than would be necessary to build a covered Apiary for ten. As the ventilators of the hive open into this Protector, the bees are, in Summer, supplied with a cool and refreshing atmosphere, as closely as possible resembling that which they find in a forest home; while in Winter, the external entrances of the hives may be safely closed, and they will receive a supply of air remarkably uniform and never much below the freezing point. As the hives themselves are double, no frost can penetrate through them, and thus their interior will almost always be perfectly dry. When the weather suddenly moderates, and bees in the common hives fly out, and are lost on the snow, those arranged in the manner described, will not know that any change has taken place, but will remain quiet in their winter quarters, unless the weather is so warm that their owner judges it safe to open the entrances, so that the warmth may penetrate their hives, and tempt them to fly, and discharge their fæces. Let it be remembered that the object of this arrangement is not to *warm up* the hives by *artificial heat*; but merely to enable the bees to retain to the utmost their own animal heat, to secure the advantages set forth in this Chapter on Protection. Once or twice during the Winter, the blocks which regulate the entrances to my hives should be removed, and as the frames are kept about half an inch from the bottom-board, by means of a stick or wire, all the dead bees and filth may, in a few moments, be removed: or as the entrance of the hives by removing the blocks, may be so enlarged as to offer no obstruction to its introduction or removal, an old newspaper can be kept on the bottom-board, and drawn out from time to time, with all its contents.

A movable board of the same thickness and length with the bottom-boards of the hive and about six inches wide, separates the hives from each other, as they stand upon the Protector.

I have made numerous observations upon the temperature of a Protector made substantially on the plan described, and find that it is wonderfully uniform. The lowest range of the thermometer during the months of January and February, 1853, in the Protector, was 28°; in the open air, 14° below zero; the highest in the Protector 32°; in the open air 56°. It will thus be seen that while the thermometer out of doors had a range of 70°, in the Protector it had a range of only 4°. While bees in common hives during some warm days flew out and perished in large numbers on the snow; the bees over the Protector were perfectly quiet. To this arrangement I attach an importance second only to my movable frames, and believe that combined with doubled hives, it removes the chief obstacle to the successful cultivation of bees in cold latitudes.[14] In the coldest regions where bees can find supplies in Summer, they may during a Winter that lasts from November to May, and during which the mercury congeals, be kept as comfortable as in climates which seem much more propitious for their cultivation. The more snow the better, as it serves more the effectually to exclude the cold from the Protector. However long and dreary the Winter, the bees in their comfortable quarters feel none of its injurious influences; and actually consume less, than those which are kept where the winters are short, and so mild that the bees are often tempted to fly, and are in a state of almost continual excitement. It is in precisely such latitudes, in Poland and Russia, that bees are kept in the largest numbers, and with the most extraordinary success. In the chapter on Pasturage, I shall show that some of the coldest places in New England, and the Middle States, are among the most favored spots for obtaining the largest supplies of the very purest honey.

Having thoroughly tested the practicability of affording the bees by my Protector, complete protection against heat and cold, at a very small expense, and in a way which may be made highly ornamental, the proper steps will be taken to secure a patent right for the same; although no extra charge will be made for this, or for any other subsequent improvement, to those who purchase the right to use my hive.

CHAPTER IX. VENTILATION OF THE HIVE.

If a populous hive is examined on a warm Summer day, a considerable number of bees will be found standing on the alighting board, with their heads turned towards the entrance, the extremity of their bodies slightly elevated, and their wings in such rapid motion that they are almost as indistinct as the spokes of a wheel, in swift rotation on its axis. A brisk current of air may be felt proceeding from the hive, and if a small piece of down be suspended by a thread, it will be blown out from one part of the entrance, and drawn in at another. What are these bees expecting to accomplish, that they appear so deeply absorbed in their fanning occupation, while busy numbers are constantly crowding in and out of the hive? and what is the meaning of this double current of air? To Huber, we owe the first satisfactory explanation of these curious phenomena. These bees plying their rapid wings in such a singular attitude, are performing the important business of *ventilating* the hive; and this double current is composed of pure air rushing in at one part, to supply the place of the foul air forced out at another. By a series of the most careful and beautiful experiments, Huber ascertained that the air of a crowded hive is almost, if not quite, as pure as the atmosphere by which it is surrounded. Now, as the entrance to such a hive, is often, (more especially in a state of nature,) very small, the interior air cannot be renewed without resort to some artificial means. If a lamp is put into a close vessel with only one small orifice, it will soon exhaust all the oxygen, and go out. If another small orifice is made, the same result will follow; but if by some device a current of air is drawn out from one, an equal current will force its way into the other, and the lamp will burn until the oil is exhausted.

It is precisely on this principle, of maintaining a double current by *artificial means*, that the bees ventilate their crowded habitations. A body of active ventilators stands inside of the hive, as well as outside, all with their heads turned towards the entrance, and by the rapid fanning of their wings, a current of air is blown briskly out of the hive, and an equal current drawn in. This important office is one which requires great physical exertion on the part of those to whom it is entrusted; and if their proceedings are carefully watched, it will be found that the exhausted ventilators, are, from time to time, relieved by fresh detachments. If the interior of the hive will admit of inspection, in very hot weather, large numbers of these ventilators will be found in regular files, in various parts of the hive, all busily engaged in their laborious employment. If the entrance at any time is contracted, a speedy accession will be made to the numbers, both inside and outside; and if it is closed entirely, the heat of the hive will quickly increase, the whole colony will commence a rapid vibration of their wings, and in a few moments will drop lifeless from the combs, for want of air.

It has been proved by careful experiments that pure air is necessary not only for the respiration of the mature bees, but that without it, neither the eggs can be hatched, nor the larvæ developed. A fine netting of air-vessels covers the eggs; and the cells of the larvæ are sealed over with a covering which is full of air holes. In Winter, as has been stated in the Chapter on Protection, bees, if kept in the dark, and neither too warm nor too cold, are almost dormant, and seem to require but a small allowance of air; but even under such circumstances, they cannot live entirely without air; and if they are excited by being exposed to atmospheric changes, or by being disturbed, a very loud humming may be heard in the interior of their hives, and they need quite as much air as in warm weather.

If at any time, by moving their hives, or in any other way, bees are greatly disturbed, it will be unsafe to confine them, especially in warm weather, unless a very free admission of air is given to them, and even then, the air ought to be admitted above, as well as below the mass of bees, or the ventilators may become clogged with dead bees, and the swarm may perish. Under close confinement, the bees become excessively heated, and the combs are often melted down. When bees are confined to a close atmosphere, especially if dampness is added to its injurious influences, they are sure to become diseased; and large numbers, if not the whole colony, perish from dysentery. Is it not under circumstances precisely similar, that cholera and dysentery prove most fatal to human beings? How often do the filthy, damp and unventilated abodes of the abject poor, become perfect lazar-houses to their wretched inmates?

I examined, last Summer, the bees of a new swarm which had been suffocated for want of air, and found their bodies distended with a yellow and noisome substance, just as though they had perished from dysentery. A few were still alive, and instead of honey, their bodies were filled with this same disgusting fluid; though the bees had not been shut up, more than two hours.

In a medical point of view, I consider these facts as highly interesting; showing as they do, under what circumstances, and how speedily, disease may be produced.

In very hot weather, if thin hives are exposed to the sun's rays, the bees are excessively annoyed by the intense heat, and have recourse to the most powerful ventilation, not merely to keep the air of the hive pure, but to carry off, as much as possible, its internal warmth. They often leave the interior of the hive, almost in a body, and in thick masses, cluster on the outside, not simply to escape the close heat within, but to guard their combs against the danger of being dissolved. At such times they are particularly careful not to cluster on the combs containing sealed honey; for as most of these combs have not been lined with the cocoons of the larvæ, they are, for this reason, as well as on account of the extra amount of wax used for their covers, much more liable to be melted, than the breeding cells.

Apiarians have often noticed the fact, that as a general thing, the bees leave the honey cells almost entirely bare, as soon as they have sealed them over; but it seems to have escaped their observation, that in hot weather, there is often an absolute necessity for such a course. In cool weather, on the contrary, the bees may often be found clustered among the sealed honey-combs, because there is then no danger of their melting down.

Few things in the range of their wonderful instincts, are so well fitted to impress the mind with their admirable sagacity, as the truly scientific device, by which these wise little insects ventilate their dwellings. I was on the point of saying that it was almost like human-reason, when the painful and mortifying reflection presented itself to my mind that in respect to ventilation, the bee is immensely in advance of the great mass of those who consider themselves as rational beings. It has, to be sure, no ability to make an elaborate analysis of the chemical constituents of the atmosphere, and to decide how large a proportion of oxygen is essential to the support of life, and how rapidly the process of breathing converts this important element into a deadly poison. It has not, like Leibig, been able to demonstrate that God has set the animal and vegetable world, the one over against the other; so that the carbonic acid produced by the breathing of the one, furnishes the aliment of the other; which, in turn,

gives out its oxygen for the support of animal life; and that, in this wonderful manner, God has provided that the atmosphere shall, through all
ages, be as pure as when it first came from His creating hand. But shame
upon us! that with all our intelligence, the most of us live as though pure
air was of little or no importance; while the bee ventilates with a scientific
precision and thoroughness, that puts to the blush our criminal neglect.

To this it may be replied that ventilation in our case, cannot be
had, without considerable expense. Can it be had for nothing, by the
industrious bees? Those busy insects, which are so indefatigably plying
their wings, are not engaged in idle amusement; nor might they, as some
would-be utilitarian may imagine, be better employed in gathering honey,
or in superintending some other department in the economy of the
hive. They are at great expense of time and labor, supplying the rest of
the colony with pure air, so conducive in every way, to their health and
prosperity.

I trust that I shall be permitted to digress, for a short time, from
bees to men, and that the remarks which I shall offer on the subject
of ventilation in human dwellings, may make a deeper impression, in
connection with the wise arrangements of the bee, than they would, if
presented in the shape of a mere scientific discussion; and that some
who have been in the habit of considering all air, except in the particular
of temperature, as about alike, may be thoroughly convinced of their
mistake.

Recent statistics prove that consumption and its kindred diseases
are most fearfully on the increase, in the Northern, and more especially in
the New England States; and that the general mortality of Massachusetts
exceeds that of almost every other state in the Union. In these States,
the tendency of increasing attention to manufacturing and mechanical
pursuits, is to compel a larger and larger proportion of the population to
lead an in-door life, and to breathe an atmosphere more or less vitiated,
and thus unfit for the full development of vigorous health. The importance
of pure air can hardly be over-estimated; indeed, the quality of the air we
breath, seems to exert an influence much more powerful, and hardly less
direct, than the mere quality of our food. Those who, by active exercise in
the open air, keep their lungs saturated as it were, with the pure element,
can eat almost anything with impunity; while those who breath the sorry
apology for air which is to be found in so many habitations, although
they may live upon the most nutritious diet, and avoid the least excess,

are incessantly troubled with head-ache, dyspepsia, and various mental as well as physical sufferings. Well may such persons, as they witness the healthy forms and happy faces of so many of the hardy sons of toil, exclaim with the old Latin poet,

"Oh dura messorum illia!"

It is with the human family very much as it is with the vegetable kingdom. Take a plant or tree, and shut it out from the pure air, and the invigorating light, and though you may supply it with an abundance of water and the very soil, which by the strictest chemical analysis, is found to contain all the elements that are essential to its vigorous growth, it will still be a puny thing, ready to droop, if exposed to a summer's sun, or to be prostrated by the first visitation of a winter's blast. Compare now, this wretched abortion, with an oak or maple which has grown upon the comparatively sterile mountain pasture, and whose branches, in Summer are the pleasant resort of the happy songsters, while, under its mighty shade, the panting herds drink in a refreshing coolness. In Winter it laughs at the mighty storms, which wildly toss its giant branches in the air, and which serve only to exercise the limbs of the sturdy tree, whose roots deep intertwined among its native rocks, enable it to bid defiance to anything short of a whirlwind or tornado.

To a population, who, for more than two-thirds of the year, are compelled to breathe an atmosphere heated by artificial means, the question how can this air be made, at a moderate expense, to resemble, as far as possible, the purest ether of the skies is, (or as I should rather say ought to be,) a question of the utmost interest. When open fires were used, there was no lack of pure air, whatever else might have been deficient. A capacious chimney carried up through its insatiable throat, immense volumes of air, to be replaced by the pure element, whistling in glee, through every crack, crevice and keyhole. Now the house-builder and stove-maker with but few exceptions[15] seem to have joined hands in waging a most effectual warfare against the unwelcome intruder. By labor-saving machinery, they contrive to make the one, the joints of his wood-work, and the other, those of his iron-work, tighter and tighter, and if it were possible for them to accomplish fully their manifest design, they would be able to furnish rooms almost as fatal to life as "the black hole of Calcutta." But in spite of all that they can do, the materials will shrink, and no fuel has yet been found, which will burn without any air, so that sufficient ventilation is kept up, to prevent such deadly occurrences. Still they are tolerably successful in keeping out the unfriendly element; and by the use of huge cooking-stoves with towering ovens, and other salamander contrivances, the little air that can find its way in, is almost as thoroughly cooked, as are the various delicacies destined for the table.

On reading an account of a run-away slave, who was for a considerable time, closely boxed up, a gentleman remarked that if the poor fellow had only known that a renewal of the air was necessary to the support of life, he could not have lived there an hour without suffocation: I have frequently thought that if the occupants of the rooms I have been describing, could only know as much, they would be in almost similar danger.

Bad air, one would think, is bad enough: but when it is heated and dried to an excessive degree, all its original vileness is stimulated to greater activity, and thus made doubly injurious by this new element of evil. Not only our private houses, but our churches and school-rooms, our railroad cars, and all our places of public assemblage, are, to a most lamentable degree, either unprovided with any means of ventilation, or, to a great extent, supplied with those which are so wretchedly deficient that they

"Keep the word of promise to our ear, And break it to our hope."

That ultimate degeneracy must surely follow such entire disregard of the laws of health, cannot be doubted; and those who imagine that the physical stamina of a people can be undermined, and yet that their intellectual, moral and religious health will suffer no eclipse or decay, know very little of the intimate connection between body and mind, which the Creator has seen fit to establish.

The men may, to a certain extent, resist the injurious influences of foul air; as their employments usually compel them to live much more out of doors: but alas, alas! for the poor women! In the very land where women are treated with more universal deference and respect than in any other, and where they so well deserve it, there often, no provision is made to furnish them with that great element of health, cheerfulness and beauty, heaven's pure, fresh air.

In Southern climes, where doors and windows may be safely kept open for a large part of the year, pure air is cheap enough, and can be obtained without any special effort: but in Northern latitudes, where heated air must be used for nearly three-quarters of the year, the neglect of ventilation is fast causing the health and beauty of our women to disappear. The pallid cheek, or the hectic flush, the angular form and distorted spine, the debilitated appearance of a large portion of our females, which to a stranger, would seem to indicate that they were just recovering from a long illness, all these indications of the lamentable absence of physical health, to say nothing of the anxious, care-worn faces and premature wrinkles, proclaim in sorrowful voices, our violation of God's physical laws, and the dreadful penalty with which He visits our transgressions.

Our people must, and I have no doubt that eventually they will be most thoroughly aroused to the necessity of a vital reform on this important subject. Open stoves, and cheerful grates and fire-places will again be in vogue with the mass of the people, unless some better mode of warming shall be devised, which, at less expense, shall make still more ample provision for the constant introduction of fresh air. Houses will be constructed, which, although more expensive in the first cost, will be far cheaper in the end, and by requiring a much smaller quantity of fuel to warm the air, will enable us to enjoy the luxury of breathing air which may be duly tempered, and yet be pure and invigorating. Air-tight and all other *lung-tight* stoves will be exploded, as economizing in fuel only when they allow the smallest possible change of air, and thus squandering health and endangering life.

The laws very wisely forbid the erection of wooden buildings in large cities, and in various ways, prescribe such regulations for the construction of edifices as are deemed to be essential to the public welfare; and the time cannot, I trust, be very far distant, when all public buildings erected for the accommodation of large numbers, will be required by law, to furnish a supply of fresh air, in some reasonable degree adequate to the necessities of those who are to occupy them.

I shall ask no excuse for the honest warmth of language which will appear extravagant only to those who cannot, or rather will not, see the immense importance of pure air to the highest enjoyment, not only of physical, but of mental and moral health. The man who shall succeed in convincing the mass of the people, of the truth of the views thus imperfectly presented, and whose inventive mind shall devise a cheap and efficacious way of furnishing a copious supply of pure air for our dwellings and public buildings, our steamboats and railroad cars, will be even more of a benefactor than a Jenner, or a Watt, a Fulton, or a Morse.

To return from this lengthy and yet I trust not unprofitable digression.

In the ventilation of my hive, I have endeavored, as far as possible, to meet all the necessities of the bees, under the varying circumstances to which they are exposed, in our uncertain climate, whose severe extremes of temperature impress most forcibly upon the bee-keeper, the maxim of the Mantuan Bard,

"Utraque vis pariter apibus metuenda."

"Extremes of heat or cold, alike are hurtful to the bees." In order to make artificial ventilation of any use to the great majority of bee-keepers, it must be simple, and not as in Nutt's hive, and many other labored contrivances, so complicated as to require almost as constant supervision as a hot-bed or a green-house. The very foundation of any system of ventilation should be such a construction of the hive that the bees shall need a change of air only for breathing.

In the Chapter on Protection, I have explained the construction of my hives, and of the Protector by which the bees being kept warm in Winter, and cool in Summer, do not require, as in thin hives, a very free introduction of air, in hot weather, to keep the combs from softening; or a still larger supply in Winter, to prevent them from moulding, and to dry up the moisture which runs from their icy tops and sides; and which, if suffered to remain, will often affect the bees with dysentery, or as it is sometimes called, "the rot." The intelligent Apiarian will perceive that I thus imitate the natural habitation of the bees in the recesses of a hollow tree in the forest, where they feel neither the extremes of heat nor cold, and where through the efficacy of their ventilating powers, a very small opening admits all the air which is necessary for respiration.

In the Chapter on the Requisites of a good hive, I have spoken of the importance of furnishing ventilation, independently of the entrance. By such an arrangement, I am able to improve upon the method which the bees are compelled to adopt in a state of nature. As they have no means of admitting air by wire-cloth, and at the same time, of effectually excluding all intruders, they are obliged in very hot weather, and in a very crowded state of their dwellings, to employ a larger force in the laborious business of ventilation, than would otherwise be necessary; while in Winter, they have no means of admitting air which is only moderately cool. I can keep the entrance so small, that only a single bee can go in at once, or I can, if circumstances require, entirely close it, and yet the bees need not suffer for the want of air. In all ordinary cases, the ventilators will admit a sufficient supply of duly tempered air from the Protector, and the bees can, at any time, increase their efficiency by their own direct agency, while yet they will, at no time, admit a strong current of chilly air, so as to endanger the life of the brood. As bees are, at all times, prone to close the ventilators with propolis, they must be placed where they can easily be removed, and cleansed, by soaking them in boiling water.

As respects ventilation from above, as well as from below, so as to allow a free current of air to pass through the hive, I am decidedly opposed to it, as in cool and windy weather, such a current often compels the bees to retire from the brood, which in this way is destroyed by a fatal chill. In thin hives, ventilation from above may be desirable in Winter, to carry off the superfluous moisture, but in properly constructed hives, standing over a Protector, there is, as has already been remarked, little or no dampness to be carried off. The construction of my hives will allow, if at all desirable, of ventilation from above; and I always make use of it, when the bees are to be shut up for any length of time, in order to be moved; as in this case, there is always a risk that the ventilators on the bottom-board may be clogged by dead bees, and the colony suffocated. As the entrance of the hive, may in a moment, be enlarged to any desirable extent, without in the least perplexing the bees, any quantity of air may be admitted, which the necessities of the bees, under any possible circumstances, may require. It may be made full 18 inches in length, but as a general rule, in Summer, in a large colony, it need not exceed six inches: while in Spring and Fall, two or three inches will suffice. In Winter, it should be entirely closed; unless in latitudes so warm, that even with the Protector, the bees cannot be kept quiet. The bee-keeper should never forget that it is almost certain destruction to a colony, to confine them when they wish to fly out. The precautions requisite to prevent robbing, will be subsequently described. In Northern latitudes, in the months of April and May, I prefer to keep the ventilators entirely closed; as the air of the Protector, at such times, like the air of a cellar in Spring, is uncomfortably cool, and has a tendency to interfere with breeding.

Note.—Since the remarks on the neglect of ventilation were put in type, my attention has been called by Hon. M. P. Wilder, of Dorchester, to an article on the same subject, in the Nov. number of the Horticulturist, for 1850, from the pen of the lamented Downing. It seems to have been written shortly after his return from Europe, and when he must have been most deeply impressed by the woeful contrast, in point of physical health between the women of America and Europe. While he speaks in just and therefore glowing terms of the virtues of our countrywomen, he says: "But in the *signs of physical health* and all that constitutes the outward aspect of the men and women of the United States, our countrymen and especially countrywomen, compare most unfavorably with all but the absolutely starving classes on the other side of the Atlantic." Close stoves he has most appropriately styled "little demons," and impure air "The favorite poison of America." His article concludes as follows:

"Pale countrymen and countrywomen rouse yourselves! Consider that God has given us an atmosphere of pure health-giving air 45 miles high, and *ventilate your houses.*"

CHAPTER X. NATURAL SWARMING, AND HIVING OF SWARMS.

The swarming of bees has been justly regarded as one of the most beautiful sights in the whole compass of rural economy. Although, for reasons which will hereafter be assigned, I prefer to rely chiefly on artificial means for the multiplication of colonies, I should be very unwilling to pass a season without participating, to some extent, in the pleasing excitement of natural swarming.

"Up mounts the chief, and to the cheated eye Ten thousand shuttles dart along the sky; As swift through æther rise the rushing swarms, Gay dancing to the beam their sun-bright forms; And each thin form, still ling'ring on the sight, Trails, as it shoots, a line of silver light. High pois'd on buoyant wing, the thoughtful queen, In gaze attentive, views the varied scene, And soon her far-fetch'd ken discerns below The light laburnum lift her polish'd brow, Wave her green leafy ringlets o'er the glade, And seem to beckon to her friendly shade. Swift as the falcon's sweep, the monarch bends Her flight abrupt; the following host descends. Round the fine twig, like cluster'd grapes, they close In thickening wreaths, and court a short repose." *Evans.*

The swarming of bees, by making provision for the constant multiplication of colonies, was undoubtedly intended both to guard the insect against the possibility of extinction, and to make its labors in the highest degree useful to man. The laws of reproduction in those insects which do not live in regular colonies, are such as to secure an ample increase of numbers. The same is true in the case of hornets, wasps and humble-bees which live in colonies only during the warm weather. In the Fall of the year, all the males perish, while the impregnated females retreat into winter quarters and remain dormant, until the warm weather restores them to activity, and each one becomes the mother of a new family.

The honey bee differs from all these insects, in being compelled, by the laws of its physical organization, to live in communities, during the entire year. The balmy breezes of Spring will quickly thaw out the frozen veins of a torpid Wasp; but the bee is incapable of enduring even a moderate degree of cold: a temperature as low as 50° speedily chills it, and it would be quite as easy to recall to life the stiffened corpses in the charnel house of the Convent of the Great St. Bernard, as to restore to animation, a frozen bee. In cool weather, they must therefore associate in large numbers, in order to maintain the animal heat which is necessary to their preservation; and the formation of new colonies, after the manner of wasps and hornets, is clearly impossible. If the young queens left the parent stock in Summer, and were able, like the mother-wasps, to lay the foundations of a new colony, they could not maintain the warmth requisite for the development of their young, even if they were able, without any baskets on their thighs, to gather bee-bread for their support. If all these difficulties were surmounted, they would still be unable to amass any treasures for our use, or even to lay up the stores requisite for their own preservation.

How admirably are all these difficulties obviated by the present arrangement! Their domicile is well supplied with all the materials for the rearing of brood, and long before any of the insects which depend upon the heat of the sun, are able to commence breeding, the bees have added thousands in the full vigor of youth to their already numerous population. They are thus able to send off in season, colonies sufficiently powerful to take advantage of the honey-harvest, and provision the new hive against the approach of Winter. From these considerations, it is very evident that swarming, so far from being, as some Apiarians have considered it, a forced or unnatural event, is one, which in a state of nature, could not possibly be dispensed with.

Let us now inquire under what circumstances it ordinarily takes place.

The time when swarms may be expected, depends of course, upon climate, season, and the strength of the stocks. In the Northern and Middle States, bees seldom swarm before the latter part of May; and June may be considered as the great swarming month. The importance of having powerful swarms early in the season, will be discussed in another place.

In the Spring, as soon as a hive well filled with comb and bees, becomes too much crowded to accommodate its teeming population, the bees begin the necessary preparations for emigration. A number of royal cells are commenced about the time that the drones first make their appearance; and by the time that the young queens arrive at maturity, the drones are always found in the greatest abundance. The first swarm is invariably led off by the old queen, unless she has previously died from accident or disease, in which case, it is accompanied by one of the young queens reared to supply her loss. The old mother leaves soon after the royal cells are sealed over, unless delayed by unfavorable weather. There are no signs from which the Apiarian can, with certainty, predict the issue of a first swarm. I devoted annually, much attention to this point, vainly hoping to discover some infallible indications of first swarming; until taught by further reflection that, from the very nature of the case, there can be no such indications. The bees, from an unfavorable state of the weather, or the failure of the blossoms to yield an abundant supply of honey, often change their minds, and refuse to swarm, even after all their preparations have been completed. Nay more, they sometimes send out no new colonies that season, when a sudden change of weather has interrupted them on the very day when they were intending to emigrate, and after they had taken a full supply of honey for their journey.

If on a fair, warm day in the swarming season, but few bees leave a strong hive, while other colonies are busily at work, we may, unless the weather suddenly prove unfavorable, look with great confidence for a swarm. As the old queens, which accompany the first swarm, are heavy with eggs, and fly with considerable difficulty, they are shy of venturing out, except on fair, still days. If the weather is very sultry, a swarm will sometimes issue as early as 7 o'clock in the morning; but from 10 to 2, is the usual time, and the majority of swarms come off from 11 to 1. Occasionally, a swarm will venture out as late as 5 P. M. An old queen is seldom guilty of such a piece of indiscretion.

I have in repeated instances witnessed the whole process of swarming, in my observing hives. On the day fixed for their departure, the queen appears to be very restless, and instead of depositing her eggs in the cells, she travels over the combs, and communicates her agitation to the whole colony. The emigrating bees fill themselves with honey, some time before their departure: in one instance, I noticed them laying in their supplies, more than two hours before they left. A short time before the swarm rises, a few bees may generally be seen, sporting in the air, with their heads turned always to the hive, occasionally flying in and out, as though they were impatient for the important event to take place. At length, a very violent agitation commences in the hive: the bees appear almost frantic, whirling around in a circle, which continually enlarges, like the circles made by a stone thrown into still water, until at last the whole hive is in a state of the greatest ferment, and the bees rush impetuously to the entrance, and pour forth in one steady stream. Not a bee looks behind, but each one pushes straight ahead, as though flying "for dear life," or urged on by some invisible power, in its headlong career. The queen often does not come out, until a large number have left, and she is frequently so heavy, from the large number of eggs in her ovaries, that she falls to the ground, incapable of rising with the colony into the air.

The bees are very soon aware of her absence, and a most interesting scene may now be witnessed. A diligent search is immediately made for their missing mother; the swarm scatters in all directions, and I have frequently seen the leaves of the adjoining trees and bushes, almost as thickly covered with the anxious explorers, as they are with drops of rain after a copious shower. If she cannot be found, they return to the old hive, though occasionally they attempt to enter some other hive, or join themselves to another swarm if any is still unhived.

The ringing of bells, and beating of kettles and frying-pans, is one of the good old ways more honored by the breach than the observance; it may answer a very good purpose in amusing the children, but I believe that as far as the bees are concerned, it is all time thrown away; and that it is not a whit more efficacious than the custom practiced by some savage tribes, who, when the sun is eclipsed, imagining that it has been swallowed by an enormous dragon, resort to the most frightful noises, to compel his snake-ship to disgorge their favorite luminary. If a swarm has selected a new home previous to their departure, no amount of *noise* will ever compel them to alight, but as soon as all the bees which compose the emigrating colony have left the hive, they fly in a direct course, or "bee-line," to the chosen spot. I have noticed that when bees are much neglected by those who pretend to take care of them, such unceremonious leave-taking is quite common; on the contrary, when proper attention is bestowed on them, it seldom occurs.

It can seldom if ever occur to those who manage their bees according to my system; as I shall show in the Chapter on Artificial Swarming. If the Apiarian perceives that his swarm instead of clustering begins to rise higher and higher in the air, and evidently means to depart, not a moment is to be lost: instead of empty noises, he must resort to means much more effective to stay their vagrant propensities. Handsful of dirt cast into the air, or water thrown among them, will often so disorganize them as to compel them to alight. Of all devices for stopping them, the most original one that I have ever heard of, is to flash the sun's rays among them, by the use of a looking glass! I have never had occasion to try it, but the anonymous writer who recommends it, says that he never knew it to fail. If they are forcibly prevented from eloping, then special care must be taken or they will be almost sure soon after hiving, to leave for their selected home. The queen should be caught and confined for several days in a way which will be subsequently described. The same caution must be exercised, when new swarms abandon their hive. If the queen cannot be caught, and there is reason to dread a desertion, the bees may be carried into the cellar, and confined in total darkness, until towards sun-set of the third day after they swarmed, being supplied in the mean time with water and honey to build their combs.

If a colony decides to go, they look upon the hive in which they are put as only a temporary stopping place, and seldom trouble themselves to build any comb in it. If the hive is so constructed as to permit inspection, I can tell by a glance whether bees are disgusted with their new residence, and mean before long to clear out. They not only refuse to work with that energy so characteristic of a new swarm, but they have a peculiar look which to the experienced eye at once proclaims the fact that they are staying only upon sufferance. Their very attitude, hanging as they do with a sort of dogged or supercilious air, as though they hated even so much as to touch their detested abode, is equivalent to an open proclamation that they mean to be off. My numerous experiments in attempting from the moment of hiving, to make the bees work in observing hives exposed to the full light of day, instead of keeping them as I now do in darkness for several days, have made me quite familiar with all their graceless, do-nothing proceedings before their departure. Bees sometimes abandon their hives very early in the Spring, or late in Summer or Fall. They exhibit all the appearance of natural swarming; but they leave, not because the population is crowded, but because it is either so small, or the hive so destitute of supplies, that they are discouraged or driven to desperation. I once knew a colony to leave the hive under such circumstances, on a spring like day in December! They seem to have a presentiment that they must perish if they stay, and instead of awaiting the sure approach of famine, they sally out to see if something cannot be done to better their condition.

At first sight, it seems strange that so provident an insect should not always select a suitable domicile before venturing on so important a step as to abandon the old home. Often before they are safely housed again, they are exposed to powerful winds and drenching rains, which beat down and destroy many of their number.

I solve this problem in the economy of the bee, in the same manner that I have solved so many others, by considering in what way, this arrangement conduces to the advantage of man.

The honey-bee would have been of comparatively little service to him, if instead of tarrying until he had sufficient time to establish them in a hive in which to labor for him, their instinct impelled them to decamp, without any delay, from the restraints of domestication. In this, as in many other things, we see that what on a superficial view, appeared to be a very obvious imperfection, proves, on closer examination, to be a special contrivance to answer important ends.

To return to our new swarm. The queen sometimes alights first, and sometimes joins the cluster after it has commenced forming. It is a very rare thing for the bees ever to cluster, unless the queen is with them; and when they do, and yet afterwards disperse, I believe that usually the queen, after first rising with them, has been lost by falling into some spot where she is unnoticed by the bees. In two instances, I performed the following interesting experiment.

Perceiving a hive in the very act of swarming, I contracted the entrance so as to secure the queen when she made her appearance. In each case, at least one third of the bees came out, before the queen presented herself to join them. When I perceived that the swarm had given up their search for her, and were beginning to return to the parent hive, I placed her, with her wings clipped, on the limb of a small evergreen tree: she crawled to the very top of the limb, as if for the purpose of making herself as conspicuous as possible. A few bees noticed her, and instead of alighting, darted rapidly away; in a few seconds, the whole colony were apprised of her presence, flew in a dense cloud to the spot, and commenced quietly clustering around her. I have often noticed the surprising rapidity with which bees communicate with each other, while on the wing. Telegraphic signals are hardly more instantaneous. (See Chapter on the Loss of the Queen.)

That bees send out scouts to seek a suitable abode, it seems to me, can admit of no serious question. Swarms have been traced to their new home, either in their flight directly from their hive, or from the place where they have clustered; and it is evident, that in such instances, they have pursued the most direct course. Now such a precision of flight to a "*terra incognita*," an unknown home, would plainly be impossible, if some of their number had not previously selected the spot, so as to be competent to act as guides to the rest. The sight of the bees for distant objects, is wonderfully acute, and after rising to a sufficient elevation, they can see the prominent objects in the vicinity of their intended abode, even although they may be several miles distant. Whether the bees send out their scouts *before* or *after* swarming, may admit of more question. In cases where the colony flies without alighting, to its new home, they are unquestionably dispatched before swarming. If this were their usual course, then we should naturally expect all the colonies to take the same speedy departure. Or if, for the convenience of the queen over fatigued by the excitement of swarming, or for any other reason, they should see fit to cluster, then we should expect that only a transient tarrying would be allowed. Instead of this, they often remain until the next day, and instances of a more protracted delay are not unfrequent. The cases which occur, of bees stopping in their flight, and clustering again on any convenient object, are not inconsistent with this view of the subject; for if the weather is hot, and the sun shines directly upon them, they will often leave before they have found a suitable habitation; and even when they are on the way to their new home, the queen being heavy with eggs, and unaccustomed to fly, is sometimes from weariness, compelled to alight, and her colony clusters around her. Queens, under such circumstances, sometimes seem unwilling to entrust themselves again to their wings, and the poor bees attempt to lay the foundations of their colony, on fence rails, hay-stacks, or other most unsuitable places.

I have been informed by Mr. Henry M. Zollickoffer of Philadelphia, a very intelligent and reliable observer, that he knew a swarm to settle on a willow tree in that city, in a lot owned by the Pennsylvania Hospital; it remained there for sometime, and the boys pelted it with stones, to get possession of its comb and honey.

The absolute necessity for scouts or explorers, is evident from all the facts in the case, unless we admit that bees have the faculty of flying in an air-line to a hollow tree, or some suitable abode which they have never seen, though they cannot find their hive, if, in their absence, it is moved only a few rods from its former position.

These obvious considerations are abundantly confirmed by the repeated instances in which a few bees have been noticed prying very inquisitively into a hole in a hollow tree or the cornice of a building, and have been succeeded, before long, by a whole colony. The importance of these remarks will be more obvious, when I come to discuss the proper mode of hiving bees.

Having described the common method of procedure pursued by the new swarm, when left without interference to their natural instincts, it is time to return to the parent stock from which they emigrated.

In witnessing the immense number which have abandoned it, we might naturally suppose that it must be almost entirely depopulated. It is sometimes asserted that as bees swarm in the pleasantest part of the day, the population is replenished by the return of large numbers of workers that were absent in the fields; this, however, can seldom be the case, as it is rare for many bees to be absent from the hive at the time of swarming.

To those who limit the fertility of the queen to 200, or at most 400 eggs per day, the rapid replenishing of the hive after swarming, must ever be a problem incapable of solution; but to those who have ocular demonstration that she can lay from one to three thousand eggs a day, it is no mystery at all. A sufficient number of bees is always left behind, to carry on the domestic operations of the hive, and as the old queen departs only when the population of the hive is super-abundant; and when thousands of young bees are hatching daily, and often 30,000 or more, are rapidly maturing, in a short time the hive is almost as populous as it was before swarming. Those who assert that the new colony is composed of young bees which have been forced to emigrate by the older ones, have certainly failed to use their eyes to much advantage, or they would have seen, in hiving a new swarm, that it is composed of both young and old; some, having wings ragged from hard work, while others are evidently quite young. After the tumult of swarming is entirely over, not a bee that did not participate in it, seeks afterwards to join the new colony, and not one that did, seeks to return. What determines some to go, and others to stay, we have no certain means of knowing.

How wonderfully abiding the impression made upon an insect, which in a moment causes it to lose all its strong affection for the old home in which it was bred, and which it has entered, perhaps hundreds of times; so that when established in another hive, though only a few feet distant, it never afterwards pays the slightest attention to its former abode! Often, when the hive into which the new swarm is put, is not removed from the place where the bees were hived, until some have gone to the fields, on their return, they fly for hours, in ceaseless circles about the spot where the missing hive stood. I have often known them to continue the vain search for their companions until they have, at length, dropped down from utter exhaustion, and perished in close proximity to their old homes!

It has been already stated that the old queen, if the weather is favorable, generally leaves about the time that the young queens are sealed over, to be changed into nymphs. In about eight days more, one of these queens hatches, and the question must now be decided whether any more colonies are to be sent out that season, or not. If the hive is well filled with bees, and the season in all respects promising, this question is generally decided in the affirmative; although colonies often refuse to swarm more than once when they are very strong, and when we can assign no reason for such a course; and they sometimes swarm repeatedly, to the utter ruin of both the old stock, and the after-swarms.

If the bees decide to swarm again, the first hatched queen is allowed to have her own way. She rushes immediately to the cells of her sisters, and, (as was described in the Chapter on Physiology,) stings them to death. From some observations that I have made, I am inclined to think that the other bees aid her in this murderous transaction: they certainly tear open the cradles of the slaughtered innocents, and remove them from the cells. Their dead bodies may often be found on the ground in front of the hive.

When a queen has emerged in the natural way from her cell, the bees usually nibble away the now useless abode, until only a small acorn cup remains; but when by violence she has met with an untimely end, they take down entirely the whole of the cell. By counting these acorn-cups, it can always be ascertained how many young queens have hatched in a hive.

Before the queens emerge from their cells, a fluttering sound is frequently heard, which is caused by the rapid motion of their wings, and which must not be confounded with the piping notes which will soon be described. If the bees of the parent stock decide to swarm again, the first hatched queen is prevented from killing the others. A strong guard is kept over their cells, and as often as she approaches them with murderous intent, she is bitten, or otherwise rudely treated, and given to understand by the most uncourtier-like demonstrations, that she cannot, in all things, do just as she pleases.

When thus repulsed, like men and women who cannot have their own way, she is highly offended and utters an angry sound, given forth in a quick succession of notes, and which sounds not unlike the rapid utterance of the words, "peep, peep." I have frequently, by holding a queen in the closed hand, caused her to make the same noise. To this angry note, one or more of the queens still unhatched, will respond, in a somewhat hoarser key, just as chicken-cocks, by crowing, bid defiance to each other. These sounds are entirely unlike the usual steady hum of the bees, and when heard, are the almost infallible indications that a second swarm will soon issue. They are occasionally so loud that they may be heard at some distance from the hive.

About a week after first swarming, the Apiarian should, early in the morning or at evening, when the bees are still, place his ear against the hive, and he will, if the queens are piping, readily recognize their peculiar sounds. If their notes are not heard, at the very latest, sixteen days after the departure of the first swarm, by which time the young queens are mature, even if the first colony left as soon as the eggs were deposited in the royal cells, it is an infallible indication that the first hatched queen is without rivals in the hive, and that swarming is over, in that stock, for the season.

The second swarm usually issues on the second or third day after this sound is heard: although I have known them to delay coming out, until the fifth day, in consequence of a very unfavorable state of the weather. Occasionally, the weather is so unfavorable, that the bees permit the oldest queen to kill the others, and refuse to swarm again. This is a rare occurrence, as the young queens, unlike the old ones, do not appear to be very particular about the weather, and sometimes venture out, not merely when it is cloudy, but even when rain is falling. On this account, if a very close watch is not kept, they are often lost. As piping ordinarily commences about eight or nine days after first swarming, the second swarm generally issues ten or twelve days after the first. It has been known to issue as early as the third day after the first, and as late as the seventeenth. Such cases, however, are of rare occurrence. It frequently happens in the agitation of swarming, that several of the young queens emerge from their cells at the same time, and accompany the colony: when this is the case, the bees often alight in two or more separate clusters. Young queens not having their ovaries burdened with eggs, are much more quick on the wing, than old ones, and fly frequently much farther from the parent stock, before they alight; though I never knew a second swarm to depart to the woods without clustering at all. After the departure of a second swarm, the oldest of the remaining queens leaves her cell; and if another swarm is to be sent forth, piping will still be heard, and so before the issue of each swarm after the first. I once had five stocks issue from one swarm, and they all came out in about two weeks. In warm latitudes more than twice this number of swarms have been known to issue in one season from a single stock. The third swarm commonly makes its appearance on the second or third day after the second swarm, and the others, at intervals of about a day.

After-swarms, or casts, (these names are given to all swarms after the first,) reduce very seriously the strength of the parent stock; for after the departure of the old queen, no more eggs are deposited in the cells, until all swarming is over. It is a very wise arrangement that the second swarm does not ordinarily issue until all the eggs left by the first queen are hatched, and the young fed and sealed over, so as to require no further care. The departure of the second swarm earlier than this, would leave too few laborers to attend to the wants of the young bees. As it is, if the weather after swarming, suddenly becomes chilly, and the hives are thin and admit too much air, the bees are too much reduced in numbers, to maintain the heat requisite for the proper development of the brood, and numbers are destroyed.

In the Chapter on Artificial Swarming, I shall discuss the effect of too frequent swarming, on the profits of the Apiary. If the bee-keeper desires to have no casts, he can, by the use of my hives, very easily, prevent their issue. As soon as the first swarm is hived, the parent stock may be opened, and all the queen cells except one removed. How much better this is, than to attempt to return the after-swarms to the parent hive, can only be appreciated by one who has thoroughly tried both plans. If the Apiarian desires the most rapid multiplication of colonies possible, where natural swarming is relied on, full directions will be furnished, in the sequel, for building up all after-swarms, however small, into vigorous stocks. It will be remembered that both the parent stock from which the swarm issues, and all the colonies except the first, have a young queen. These queens never leave the hive for impregnation, until after they have been established as the acknowledged heads of independent families. They generally go out for this purpose, the first pleasant day after they are thus acknowledged, early in the afternoon, at which hour the drones are flying in the greatest numbers. On first leaving their hive, they always fly with their heads turned towards it, and enter and depart often several times before they finally soar up into the air. Such precautions on the part of a young queen, are highly necessary that she may not mistake her own hive on her return, and lose her life by attempting to enter that of another colony. Mistakes of this kind are frequently made when the hives stand near, and closely resemble each other, and are fatal, not only to the queen, but to her whole colony. In the new hive there is no brood at all, and in the old one it is too far advanced towards maturity to answer for raising new queens. Such calamities, in my hive, admit of a very easy remedy, as I shall show in the Chapter on the Loss of the Queen.

To guard the young queen against such frequent mistakes, I paint the covered fronts of my hives, with the alighting boards, and blocks guarding the entrance, of different colors. This answers the same purpose as to paint the whole surface of the boxes, some of one color, and some of another. The only proper color for a hive when exposed to the weather, is a perfect white; any shade of color will absorb the heat of the sun, so as to warp the wood-work of the hive, besides exposing the bees to a pent and suffocating heat.

When a young queen leaves the hive for the purpose above mentioned, the bees, on missing her, are often filled with alarm, and rush from the hive, just as though they were intending to swarm. Their agitation soon calms down, if she returns to them in safety. I shall give through the medium of the Latin tongue, some statements which are important only to the scientific naturalist, and entomologist.

Post coitum fucus statim perit. Penis ejectio, ut ego comperi, lenem compressionem fuci ventris, consequitur; et fucus extemplo similis fulmine tacto, moritur. Dominus Huber sæpe videbat fuci organum post congressum, in corpore feminæ hæsisse. Vidi semel tam firme inhærens, ut nisi disruptione reginæ ventris, non possim divellere.

The queen commences laying eggs, about two days after impregnation, and for the first season, lays none but the eggs of workers; no males being needed in colonies which will throw no swarm till another season. It is seldom until after she has commenced replenishing the cells with eggs, that she is treated with any special attention by the bees; although if deprived of her before this time, they show, by their despair, that they thoroughly comprehended her vast importance to their welfare.

I shall now give such practical directions for the easy hiving of swarms, as will, I trust, greatly facilitate the whole operation, not merely to the novice, but even to many experienced bee-keepers; and I shall try to make these directions sufficiently minute, to guide those who having never seen a swarm hived, are very apt to imagine that the process must be a formidable one, instead of being, as it usually is to those who are fond of bees, a most delightful entertainment. Experience in this, as in other things, will speedily give the requisite skill and confidence; and the cry of "the bees are swarming," will soon be hailed with greater pleasure than an invitation to the most sumptuous banquet.

The hives for the new swarms should all be in readiness before the swarming season begins, and should be painted long enough beforehand, to have the paint most thoroughly dried. The smell of fresh paint is well known to be exceedingly injurious to human beings, and is such an abomination to the bees, that they will often desert a new hive sooner than put up with it. If the hives cannot be painted in ample season, then such paints should be used, as contain no white lead, and they should be mixed in such a manner as to dry as quickly as possible. Thin hives ought never to stand in the sun, and then, when heated to an insufferable degree, be used for a new swarm. Bees often refuse to enter such hives at all, and at best, are very slow in taking possession of them. It should be borne in mind, that bees, when they swarm, are greatly excited, and unnaturally heated. The temperature of the hive, at the moment of swarming, rises very suddenly, and many of the bees are often drenched with such a profuse perspiration that they are unable to take wing and join the departing colony. The attempt to make bees enter a heated hive in a blazing sun, is as irrational as it would be to try to force a panting crowd of human beings into the suffocating atmosphere of a close garret. If bees are to be put in hives through which the heat of the sun can penetrate, the process should be accomplished in the shade, or if this cannot conveniently be done, the hive should be covered with a sheet, or shaded with leafy boughs. If a hive with my movable frames is used, these should all be furnished, or at least, every other one, with a small piece of worker-comb, attached to the center of the frame, with melted wax or rosin. Without such a guide comb, the bees will almost always work some of the combs out of the true direction, and this will interfere with their easy removal. A sheet of comb, not larger than five inches square, will answer for all the frames of one hive. If even so small a piece of comb as this cannot be procured, let a thin line of melted wax be drawn, lengthwise, over the middle of each frame, and let the colony be examined, on the second day after hiving, and all the frames which contain irregular comb, be removed. This comb may be cut off, and attached so as to serve as a proper guide to the bees. The possession of six frames containing good worker comb, and wrought with perfect accuracy, may be made by the following device, to answer a most admirable end. Put them into a hive with six empty frames; first a frame with comb, then an empty one, &c. After the bees have had possession of this hive two or three days, visit them, and very politely inform them that the full frames were intended as a loan, and not as a gift; and that having served to set them an example how they should work, you must now have them to teach other young swarms the same useful lesson; and that the new combs which they have built with such admirable regularity, are beautiful patterns for the empty ones which you must give them. In this way, the same combs may be made to answer for many successive swarms.

Drone combs should never be attached to the frames as a guide, unless it is desired to have the bees follow the pattern, and build large ranges of drone comb, to breed a vast horde of useless consumers. Such comb, if white, may be used to great advantage in the surplus honey-boxes; if old and discolored, it should be melted for wax. I am now engaged in a course of experiments, which I hope, will enable me to dispense with the necessity of guide-combs for my frames, as they are sometimes difficult to be procured by those who have just commenced bee-keeping. As a general thing, however, every one, after a few weeks' experience, may have enough and to spare, for such purposes. Every piece of good worker-comb, if large enough to be attached to a frame, should be used both for its intrinsic value, and because bees are so wonderfully pleased when they find such unexpected treasures in a hive, that they will seldom desert it. A new swarm has been known to take possession of an old hive without any occupants, but well stored with comb. Though dozens of empty hives may be in the Apiary, they never unless under such circumstances, enter a hive, of their own accord. It might seem as though an instinct impelling them to do so, would have been a most admirable one, and so doubtless, it may seem to some that it would have been much better for man, if the earth had only brought forth spontaneously all things requisite for the support of man and beast, without any necessity for the sweat of the brow. The first and last frames in my hive, are placed about a quarter of an inch from the ends, and the others just half an inch apart. When first put in, it will be advisable to attach them slightly with a very little glue or melted wax, to keep them in their places, until they are fastened with propolis, by the bees. The rubbing of hives with various kinds of herbs or washes, has always seemed to me, useless, and often positively injurious. There ought always to be some small trees near the hives, on which the swarms can cluster, and from which they can be easily gathered. If there are none, limbs of trees about six feet high, (evergreens are best,) may be fastened into the ground, a few rods in front of the hives, and they will answer a very good temporary purpose. It will inspire the inexperienced Apiarian with much greater confidence, to remember that almost all the bees in a swarm, have filled themselves with honey, before leaving the parent stock, and are therefore in a very peaceable mood. If he is at all timid, or liable, as some are, to suffer severely from the sting of a single bee, he should, by all means, furnish himself with the protection of a bee-dress. (See Bee-Dress.)

I shall, in another place, give the best remedies for the relief of a sting. As soon as the bees have quietly clustered around their queen, preparation should be made to hive them without any unnecessary delay. The headlong haste of some Apiarians, which, by throwing them into a profuse perspiration, renders them very liable to be stung, is altogether unnecessary. The very fact that the bees have clustered, after leaving the parent stock, is almost equivalent to a certainty that they will not leave, for at least one or two hours. All convenient despatch should be used, however, lest other colonies issue before the first one is hived, and attempt to add themselves, as they frequently do, to the first swarm. The proper course to be pursued, in such a case, will be subsequently explained. If my hives are used, the entrance on the whole front must be opened, so that the bees may have every chance to enter as rapidly as possible; and a sheet must be fastened to the alighting-board, to keep the bees from being separated from each other or soiled by dirt, for a bee thoroughly covered with dust or dirt, is almost sure to perish. Unless the bees cluster at a considerable distance from the place where they are intended to be permanently stationed, the new hive which receives them may stand on the Protector in its proper place, with the sheet tacked or pinned to the alighting-board, and spread out over the mound in front of the entrance. If the common hives are used, they must generally be carried to the swarm, and propped up on the sheet, so as to give the bees a free admission. When the bees alight where they can be easily reached from the ground, the limb on which they have clustered, should, with one hand, be shaken, so that they may gently fall into a basket held under them, by the other. If the basket is sufficiently open to admit the air freely, and not so open as to allow the bees to get through the sides, it will answer all the better. The bees should now be carried very slowly to their new home, and be gently shaken, or poured out, on the sheet, in front of it. If they seem at all reluctant to enter, take up a few of them in a large spoon, (a cup will answer equally well,) and shake them close to the entrance. As they go in, they will fan with their wings, and raise a peculiar note, which communicates the joyful news that they have found a home, to the rest of their companions; and in a short time, the whole swarm will enter, and they are thus safely hived, without injury to a single bee. When bees are once shaken down on the sheet, the great mass of them are very unwilling to take wing again; for they are loaded with honey, and like heavily armed troops, they desire to march slowly and sedately to the place of encampment. If the sheet hangs in folds, or is not stretched out, so as to present an uninterrupted surface, they are often greatly confused, and take a long time to find the entrance to the hive. If it is desired to have them enter sooner than they are sometimes inclined to do, they may be gently separated, with a feather, or leafy twig, when they cluster in bunches on the sheet. On first shaking them down into the basket, multitudes will again take wing, and multitudes more will be left on the tree, but they will speedily form a line of communication with those on the sheet, and enter the hive with them; for many of them will follow the Apiarian, as he slowly carries the basket to the hive.

It sometimes happens that the queen is left on the tree: in this case, the bees will either refuse to enter the hive, or if they go in, will speedily come out, and all take wing again, to join their queen. This happens much more frequently in the case of after-swarms, whose young queens, instead of exhibiting the gravity of the old matron, are apt to be constantly flying about, and frisking in the air. When the bees cluster again on the tree, the process of hiving must be repeated.

If the Apiarian has a pair of sharp pruning-shears, and the limb on which the bees have clustered, is of no value, and so small, that it can be cut without jarring them off, this may be done, and the bees carried on it and then shaken off on the sheet.

If the bees settle too high to be easily reached, the basket should be fastened to a pole, and raised directly under the swarm; a quick motion of the basket will cause the mass of the bees to fall into it, when it may be carried to the hive, and the bees poured out from it on the sheet.

If the bees light on the trunk of a tree, or any thing from which they cannot easily be gathered in a basket, place a leafy bough over them, (it may be fastened with a gimlet,) and if they do not mount it of their own accord, a little smoke will compel them to do so. If the place is inaccessible, and this is about the worst case that occurs, they will enter a basket well shaded by cotton cloth fastened around it, and elevated so as to rest with its open top sideways to the mass of the bees. When small trees, or limbs fastened into the ground, are placed near the hives, and there are no large trees near, there will seldom be found any difficulty in hiving swarms. If two swarms light together, I advise that they should be put into one hive, and abundant room at once be given them, for storing surplus honey. This can always be readily done in my hives. Large quantities of honey are generally obtained from such stocks, if the season is favorable, and they have issued early. If it is desired to separate them, place in each of the hives which is to receive them, a comb containing brood and eggs, from which, in case of necessity, a new queen may be raised. Shake a portion of the bees in front of each hive, sprinkling them thoroughly, both before and after they are shaken out from the basket, so that they will not take wing to unite again. If possible, secure the queens, so that one may be given to each hive. If this cannot be done, the hives should be examined the next day, and if the two queens entered the same hive, one will have killed the other, and the queenless hive will be found building royal cells. It should be supplied with a sealed queen nearly mature, taken from another hive, not only to save time, but to prevent them from filling their hive with comb unfit for the rearing of workers. (See Artificial Swarming.) Of course, this cannot be done with the common hives, and if the Apiarian does not succeed in getting a queen for each hive, the queenless one will refuse to stay, and will go back to the old stock.

The old-fashioned way of hiving bees, by mounting trees, cutting and lowering down large limbs, (often to the injury of valuable trees,) and placing the hive over the bees, frequently crushing large numbers, and endangering the life of the queen, should be entirely abandoned. A swarm may be hived in the proper way with far less risk and trouble, and in much less time. In large Apiaries managed on the swarming plan, where a number of swarms come out on the same day, and there is constant danger of their mixing,[16] the speedy hiving of swarms is an object of great importance. If the new hive does not stand where it is to remain for the season, it should be removed to its permanent stand as soon as the bees have entered; for if allowed to remain to be removed in the evening, or early next morning, the scouts which have left the cluster, in search of a hollow tree, will find the bees when they return, and will often entice them from the hive. There is the greater danger of this, if the bees have remained on the tree, a considerable time before they were hived. I have invariably found that swarms which abandon a suitable hive for the woods, have been hived near the spot where they clustered, and allowed to remain to be moved in the evening. If the bees swarm early in the day, they will generally begin to work in a few hours (or in less time, if they have empty comb,) and many more may be lost by returning next day to the place where they were hived, than would be lost, by removing them as soon as they have entered; in this latter case, the few that are on the wing, will generally be able to find the hive if it is slowly moved to its permanent stand.

If the Apiarian wishes to secure the queen, the bees should be shaken from the hiving basket, about a foot from the entrance to the hive, and if a careful look-out is kept, she will generally be seen as she passes over the sheet, to the entrance. Care must be taken to brush the bees back from the entrance when they press forward in such dense masses that the queen is likely to enter unnoticed. An experienced eye readily catches a glance of her peculiar form and color. She may be taken up without danger, as she never stings, unless engaged in combat with another queen. As it will sometimes happen, even to careful bee-keepers, that swarms will come off when no suitable hives are in readiness to receive them, I shall show what may be done in such an emergency. Take any old hive, box, cask, or measure, and hive the bees in it, placing them with suitable protection against the sun, where their new hive is to stand; when this is ready, they may, by a quick jerking motion, be easily shaken out on a sheet, and hived in it, just as though they were shaken from the hiving basket. If they are to remain in the temporary hive over the second day, they ought to be shaken out on a sheet, and after their comb is taken from them, allowed to enter it again, or else there will be danger of crushing the queen by the weight of the comb.

I have endeavored, even at the risk of being tedious, to give such specific directions as will qualify the novice to hive a swarm of bees, under almost any circumstances; for I know the necessity of such directions and how seldom they are to be met with, even in large treatises on Bee-Keeping. Vague or imperfect directions always fail, just at the moment that the inexperienced attempt to put them into practice.

Before leaving this subject, I will add to the directions for hiving already given, a method which I have practiced with good success.

When the situation of the bees does not admit of the basket being easily elevated to them, the bee-keeper may carry it with him to the cluster, and then after shaking the bees into it, may lower it down by a string, to an assistant standing below.

That Natural Swarming may, with suitable hives, be made highly profitable, I cannot for a moment question. As it is the most simple and obvious way of multiplying colonies, and the one which requires the least amount of knowledge or skill, it will undoubtedly, for many years at least, be the favorite method with a large number of bee-keepers. I have therefore, been careful to furnish suitable directions for its successful practice; and before I discuss the question of Artificial Increase, I shall show how it may be more profitably conducted than ever before; many of the most embarrassing difficulties in the way of its successful management being readily obviated by the use of my hives.

1. The common hives fail to furnish adequate protection in Winter, against cold, and those sudden changes to unseasonable warmth, by which bees are tempted to come out and perish in large numbers on the snow; and the colonies are thus prevented from breeding on a large scale, as early as they otherwise would. Under such circumstances, they can make no profitable use of the early honey-harvest; and they will swarm so late, if they swarm at all, as to have but little opportunity for laying up surplus honey, while often they do not gather enough even for their own use, and their owner closes the season by purchasing honey to preserve them from starvation. The way in which I give the bees that amount of protection in Winter, which conduces most powerfully to early swarming, has already been described in the Chapter on Protection.

2. Another serious objection to all the ordinary swarming hives, is the vexatious fact that if the bees swarm at all, they are liable to swarm so often as to destroy the value of both the parent stock and the after-swarms. Experienced bee-keepers obviate this difficulty, by uniting second swarms, so as to make one good colony out of two; and they return to the parent stock all swarms after the second, and even this if the season is far advanced. Such operations consume much time, and often give much

more trouble than they are worth. By removing all the queen cells but one, after the first swarm has left, second swarming in my hives will always be prevented; and by removing all but two, provision may be made for the issue of second swarms, and yet all after-swarming be prevented. The process of returning after-swarms is not only objectionable, on account of the time it requires, having often to be repeated again and again before one queen is allowed to destroy the others; but it also causes a large portion of the gathering season to be wasted; for the bees seem unwilling to work with energy, so long as the pretensions of several rival queens are unsettled.

3. Another very serious objection to Natural Swarming, as practiced with the common hives, is the inability of the Apiarian who wishes rapidly to multiply his colonies, to aid his late and small swarms, so as to build them up into vigorous stocks. The time and money which are ordinarily spent upon small colonies, are almost always thrown away; by far the larger portion of them never survive the Winter, and the majority of those that do, are so enfeebled, as to be of little or no value. If they escape being robbed by stronger stocks, or destroyed by the moth, they seldom recruit in season to swarm, and very often the feeding must be repeated, the second Fall, or they will at last perish. I doubt not that many of my readers will, from their own experience, endorse every word of these remarks, as true to the very letter. All who have ever attempted to multiply colonies by nursing and feeding small swarms, on the ordinary plans, have found it attended with nothing but loss and vexation. The more a man has of such stocks, the poorer he is: for by their weakness, they are constantly tempting his strong swarms to evil courses; so that at last, they prefer to live as far as they can, by stealing, rather than by habits of honest industry; and if the feeble colonies escape being plundered, they often become mere nurseries for raising a plentiful supply of moths, to ravage his whole Apiary.

I have already shown, in what way by the use of my hives, the smallest swarms that ever issue, may be so managed as to become powerful stocks. In the same way the Apiarian can easily strengthen all his colonies which are feeble in Spring.

4. As the loss of the young queens in the parent stock after it has swarmed, and in the after-swarms, is a very common occurrence, a hive which like mine, furnishes the means of easily remedying this misfortune, will greatly promote the success of those who practice natural swarming.

A very intelligent bee-keeper once assured me, that he must use at least one such hive in his Apiary, for this purpose, even if in other respects it possessed no superior merits.

5. Bees, as is well known, often refuse to swarm at all, and most of the swarming hives are so constructed, that proper accommodations for storing honey, cannot be furnished to the super-abundant population. Under such circumstances, they often hang for several months, in black masses on the outside of the hive; and are worse than useless, as they consume the honey which the others have gathered. In my hives, an abundance of room for storing honey can always be given them, *not all at once*, so as to prevent them from swarming, but by degrees, as their necessities require: so that if they are indisposed, for any reason to swarm, they may have suitable receptacles easily accessible, and furnished with guide comb to make them more attractive, in which to store up any amount of honey that they can possibly collect.

6. In the common hives, but little can be done to dislodge the bee-moth, when once it has gained the mastery of the bees; whereas in mine, it can be most effectually rooted out when it has made a lodgment. (See Remarks on Bee-Moth.)

7. In the common hives, nothing can be done except with great difficulty, to remove the old queen when her fertility is impaired; whereas in my hives, (as will be shown in the Chapter on Artificial Swarming,) this can easily be effected, so that an Apiary may constantly contain a stock of young queens, in the full vigor of their re-productive powers.

I trust that these remarks will convince intelligent Apiarians, that I have not spoken boastfully or at random, in asserting that natural swarming can be carried on with much greater certainty and success, by the use of my hives, than in any other way; and that they will see that many of the most perplexing embarrassments and mortifying discouragements under which they have hitherto prosecuted it, may be effectually remedied.

CHAPTER XI. ARTIFICIAL SWARMING.

The numerous efforts which have been made for the last fifty years or more, to dispense with natural swarming, plainly indicate the anxiety of Apiarians to find some better mode of increasing their colonies.

Although I am able to propagate bees by natural swarming, with a rapidity and certainty unattainable except by the complete control of all the combs in the hive, still there are difficulties in this mode of increase, inherent to the system itself, and therefore entirely incapable of being removed by any kind of hive. Before describing the various methods which I employ to increase colonies by artificial means, I shall first enumerate these difficulties, in order that each individual bee-keeper may decide for himself, in which way he can most advantageously propagate his bees.

1. The large number of swarms lost every year, is a powerful argument against natural swarming.

An eminent Apiarian has estimated that one fourth of the best swarms are lost every season! This estimate can hardly be considered too high, if all who keep bees are taken into account. While some bee-keepers are so careful that they seldom lose a swarm, the majority, either from the grossest negligence, or from necessary hindrances during the swarming season, are constantly incurring serious losses, by the flight of their bees to the woods. It is next to impossible, entirely to prevent such occurrences, if bees are allowed to swarm at all.

2. The great amount of time and labor required by natural swarming, has always been regarded as a decided objection to this mode of increase.

As soon as the swarming season begins, the Apiary must be closely watched almost every day, or some of the new swarms will be lost. If this business is entrusted to thoughtless children, or careless adults, many swarms will be lost by their neglect. It is very evident that but few persons who keep bees, can always be on hand to watch them and to hive the new swarms. But, in the height of the swarming season, if any considerable number of colonies is kept, the Apiarian, to guard against serious losses, should either be always on the spot himself, or have some one who can be entrusted with the care of his bees. Even the Sabbath cannot be observed as a day of rest; and often, instead of being able to go to the House of God, the bee-keeper is compelled to labor among his bees, as hard as on other days, or even harder. That he is as justifiable in hiving his bees on the Sabbath, as in taking care of his stock, can admit of no serious doubt; but the very liability of being called to do so, is with many, a sufficient objection against Apiarian pursuits.

The merchant, mechanic and professional man, are often so situated that they would take great interest in bees, if they were not deterred from their cultivation by inability to take care of them, during the swarming season; and they are thus debarred from a pursuit, which is intensely fascinating, not merely to the lover of Nature, but to every one possessed of an inquiring mind. No man who spends some of his leisure hours in studying the wonderful habits and instincts of bees, will ever complain that he can find nothing to fill up his time out of the range of his business, or the gratification of his appetites. Bees may be kept with great advantage, even in large cities, and those who are debarred from every other rural pursuit, may still listen to the soothing hum of the industrious bee, and harvest annually its delicious nectar.

If the Apiarian could always be on hand during the swarming season, it would still, in many instances, be exceedingly inconvenient for him to attend to his bees. How often is the farmer interrupted in the business of hay-making, by the cry that his bees are swarming; and by the time he has hived them, perhaps a shower comes up, and his hay is injured more than his swarm is worth. In this way, the keeping of a few bees, instead of a source of profit, often becomes rather an expensive luxury; and if a very large stock is kept, the difficulties and embarrassments are often most seriously increased. If the weather becomes pleasant after a succession of days unfavorable for swarming, it often happens that several swarms rise at once, and cluster together, to the great annoyance of the Apiarian; and not infrequently, in the noise and confusion, other swarms fly off, and are entirely lost. I have seen the Apiarian so perplexed and exhausted under such circumstances, as to be almost ready to wish that he had never seen a bee.

3. The managing of bees by natural swarming, must, in our country, almost entirely prevent the establishment of large Apiaries.

Even if it were possible, in this way, to multiply bees with certainty and rapidity, and without any of the perplexities which I have just described, how few persons are so situated as to be able to give almost the whole of their time in the busiest part of the year, to the management of their bees. The swarming season is with the farmer, the very busiest part of the whole year, and if he purposes to keep a large number of swarming hives, he must not only devote nearly the whole of his time, for a number of weeks, to their supervision, but at a season when labor commands the highest price, he will often be compelled to hire additional assistance.

I have long been convinced that, as a general rule, the keeping of a few colonies in swarming hives, costs more than they are worth, and that the keeping of a very large number is entirely out of the question, unless with those who are so situated that they can afford to devote their time, for about two months every year, almost entirely to their bees. The number of persons who can afford to do this must be very small; and I have seldom heard of a bee-keeper, in our country, who has an Apiary on a scale extensive enough to make bee-keeping anything more than a subordinate pursuit. Multitudes have tried to make it a large and remunerating business, but hitherto, I believe that they have nearly all been disappointed in their expectations. In such countries as Poland and Russia where labor is deplorably cheap, it may be done to great advantage; but never to any considerable extent in our own.

4. A serious objection to natural swarming, is the discouraging fact that the bees often refuse to swarm at all, and the Apiarian finds it impossible to multiply his colonies with any certainty or rapidity, even although he may find himself in all respects favorably situated for the cultivation of bees, and may be exceedingly anxious to engage in the business on a much more extensive scale.

I am acquainted with many careful bee-keepers who have managed their bees according to the most reliable information they could obtain, never destroying any of their colonies, and endeavoring to multiply them to the best of their ability, who yet have not as many stocks as they had ten years ago. Most of them would abandon the pursuit, if they looked upon bee-keeping simply in the light of dollars and cents, rather than as a source of pleasant recreation; and some do not hesitate to say that much more money has been spent, by the mass of those who have used patent hives, than they have ever realized from their bees.

It is a very simple matter to make calculations on paper, which shall seem to point out a road to wealth, almost as flattering, as a tour to the gold mines of Australia or California. Only purchase a patent bee-hive, and if it fulfills all or even a part of the promises of its sanguine inventor, a fortune must, in the course of a few years, be certainly realized; but such are the disappointments resulting from the bees refusing often to swarm at all, that if the hive could remedy all the other difficulties in the way of bee-keeping, it would still fail to answer the reasonable wishes of the experienced Apiarian. If every swarm of bees could be made to yield a profit of 20 dollars a year, and if the Apiarian could be sure of selling his new swarms at the most extravagant prices, he could not, like the growers of mulberry trees, or the breeders of fancy fowls, multiply his stocks so as to meet the demand, however extensive; but would be entirely dependent upon the whims and caprices of his bees; or rather, upon the natural laws which control their swarming.

Every practical bee-keeper is well aware of the utter uncertainty of natural swarming. Under no circumstances, can its occurrence be confidently relied on. While some stocks swarm regularly and repeatedly, others, strong in numbers and rich in stores, although the season may, in all respects, be propitious, refuse to swarm at all. Such colonies, on examination, will often be found to have taken no steps for raising young queens. In some cases, the wings of the old mother will be found defective, while in others, she is abundantly able to fly, but seems to prefer the riches of the old hive, to the risks attending the formation of a new colony. It frequently happens, in our uncertain climate, that when all the necessary preparations have been made for swarming, the weather proves unpropitious for so long a time, that the young queens coming to maturity before the old one can leave, are all destroyed. This is a very frequent occurrence, and under such circumstances, swarming is almost certain to be prevented, for that season. The young queens are frequently destroyed, even although the weather is pleasant, in consequence of some sudden and perhaps only temporary suspension of the honey-harvest; for bees seldom colonize even if all their preparations are completed, unless the flowers are yielding an abundant supply of honey.

From these and other causes which my limits will not permit me to notice, it has hitherto been found impossible, in the uncertain climate of our Northern States, to multiply colonies very rapidly, by natural swarming; and bee-keeping, on this plan, offers very poor inducements to those who are aware how little has been accomplished, even by the most enthusiastic, experienced and energetic Apiarians.

The numerous perplexities which have ever attended natural swarming, have for ages, directed the attention of practical cultivators, to the importance of devising some more reliable method of increasing their colonies. Columella, who lived about the middle of the first century of the Christian Era, and who wrote twelve books on husbandry (De re rustica,) has given directions for making artificial colonies. He says, "you must examine the hive, and view what honey-combs it has; then afterwards from the wax which contains the seeds of the young bees, you must cut away that part wherein the offspring of the royal brood is animated: for this is easy to be seen; because at the very end of the wax-works there appears, as it were, a thimble-like process (somewhat similar to an acorn,) rising higher, and having a wider cavity, than the rest of the holes, wherein the young bees of vulgar note are contained."

Hyginus, who flourished before Columella, had evidently noticed the royal jelly; for he speaks of cells larger than those of the common bees, "filled as it were with a solid substance of a *red color*, out of which the winged king is at first formed." This ancient observer must undoubtedly have seen the quince-like jelly, a portion of which is always found at the base of the royal cells, after the queens have emerged. The ancients gener-ally called the queen a king, although Aristotle says that some in his time called her the mother. Swammerdam was the first to prove by dissection that the queen is a perfect female, and the only one in the hive, and that the drone is the male.

For reasons which I shall shortly mention, the ancient methods of artificial increase appear to have met with but small success. Towards the close of the last century, a new impulse was given to the artificial produc-tion of swarms, by the discovery of Schirach, a German clergyman, that bees are able to rear a queen from worker-brood. For want, however, of a more thorough knowledge of some important principles in the economy of the bee, these efforts met with slender encouragement.

Huber, after his splendid discoveries in the physiology of the bee, perceived at once, the importance of multiplying colonies by some method more reliable than that of natural swarming. His leaf or book hive consisted of 12 frames, each an inch and a half in width; any one of which could be opened at pleasure. He recommends forming artificial swarms, by dividing one of these hives into two parts; adding to each part six empty frames. After using a Huber hive for a number of years, I became perfectly convinced that it could only be made servicable, by an adroit, experienced and fearless Apiarian. The bees fasten the frames in such a manner, with their propolis, that they cannot, except with extreme care, be opened without jarring the bees, and exciting their anger; nor can they be shut without constant danger of crushing them. Huber nowhere speaks of having multiplied colonies extensively by such hives, and although they have been in use more than sixty years, they have never been successfully employed for such a purpose. If Huber had only contrived a plan for suspending his frames, instead of folding them together like the leaves of a book, I believe that the cause of Apiarian science would have been fifty years in advance of what it now is.

Dividing hives of various kinds have been used in this country. After giving some of the best of them a thorough trial, and inventing others which somewhat resembled the Huber hive, I found that they could not possibly be made to answer any valuable end in securing artificial swarms. For a long time I felt that the plan *ought* to succeed, and it was not until I had made numerous experiments with my hive substantially as now constructed, that I ascertained the precise causes of failure.

It may be regarded as one of the laws of the bee-hive, that bees, when not in possession of a mature queen, seldom build any comb except such as being designed merely for storing honey, is *too coarse for the rearing of workers*. Until I became acquainted with the discoveries of Dzierzon, I supposed myself to be the only observer who had noticed this remarkable fact, and who had been led by it, to modify the whole system of artificial swarming. The perusal of Mr. Wagner's manuscript translation of that author, showed me that he had arrived at precisely similar results.

It may seem at first, very unaccountable that bees should go on to fill their hives with comb unfit for breeding, when the young queen will so soon require worker-cells for her eggs; but it must be borne in mind, that bees, under such circumstances, are always in an *unnatural* state. They are attempting to rear a new queen in a hive which is only partially filled with comb; whereas, if left to follow their own instincts, they never construct royal cells except in hives which are well filled with comb, for it is only in such hives that they make any preparations for swarming. It must be confessed that they do not show their ordinary sagacity in filling a hive with unsuitable comb; but if it were not for a few instances of this kind of bad management, we should perhaps, form too exalted an idea of their intelligence, and should almost fail to notice the marked distinction between reason in man, and even the most refined instincts of some of the animals by which he is surrounded.

The determination of bees, when they have no mature queen,

if they build any comb at all, to build such as is suited only for storing honey, and unfit for breeding, will show at once, the folly of attempting to multiply colonies by the dividing-hives. Even if the Apiarian has been perfectly successful in dividing a colony, and the part without a queen takes the necessary steps to supply her loss, if the bees are sufficiently numerous to build a large quantity of new comb, (and they ought to be in order to make the artificial colony of any value,) they will build this comb in such a manner that it will answer only for storing honey, while they will use the half of the hive with the old comb, for the purposes of breeding. The next year, if an attempt is made to divide this hive, one half will contain nearly all the brood and mature bees, while the other, having most of the honey, in combs unfit for breeding, the new colony formed from it will be a complete failure.

Even with a Huber hive, the plan of multiplying colonies by dividing a full hive into two parts, and adding an empty half to each, will be attended with serious difficulties; although some of them may be remedied in consequence of the hive being constructed so as to divide into many parts; the very attempt to remedy them, however, will be found to require a degree of skill and knowledge far in advance of what can be expected of the great mass of bee-keepers.

The common dividing hives, separating into two parts, can never, under any circumstances, be made of the least practical value; and the business of multiplying colonies by them, will be found far more laborious, uncertain and vexatious, than to rely on natural swarming. I do not know of a solitary practical Apiarian, who, on trial of this system, has not been compelled to abandon it, and allow the bees to swarm from his dividing hives in the old-fashioned way.

Some Apiarians have attempted to multiply their colonies by putting a piece of brood comb containing the materials for raising a new queen, into an empty hive, set in the place of a strong stock which has been removed to a new stand when thousands of its inmates were abroad in the fields. This method is still worse than the one which has just been described. In the dividing hive, the bees already had a large amount of suitable comb for breeding, while in this having next to none, they build all their combs until the queen is hatched, of a size unsuitable for rearing workers. In the first case, the queenless part of the dividing hive may have had a young queen almost mature, so that the process of building large combs would be of short continuance; for as soon as the young queen begins to lay, the bees at once commence building combs adapted to the reception of worker eggs. In some of my attempts to rear artificial swarms by moving a full stock, as described above, I have had combs built of enormous size, nearly four inches through! and these monster combs have afterwards been pieced out on their lower edge, with worker cells for the accommodation of the young queen! So uniformly do the bees with an unhatched queen, build in the way described, that I can often tell at a single glance, by seeing what kind of comb they are building, that a hive is queenless, or that having been so, they have now a fertile young queen. When a new colony is formed, by dividing the old hive, the queenless part has thousands of cells filled with brood and eggs, and young bees will be hourly hatching, for at least three weeks: and by this time, the young queen will be laying eggs, so that there will be an interval of not more than three weeks, during which no accessions will be made to the numbers of the colony. But when a new swarm is formed by moving, not an egg will be deposited for nearly three weeks; and not a bee will be hatched for nearly six weeks; and during all this time, the colony will rapidly decrease, until by the time that the progeny of the young queen begins to emerge from their cells, the number of bees in the new hive will be so small, that it would be of no value, even if its combs were of the best construction.

Every observing bee-keeper must have noticed how rapidly even a powerful swarm diminishes in number, for the first three weeks after it has been hived. In many cases, before the young begin to hatch, it does not contain one half its original number; so very great is the mortality of bees during the height of the working season.

I have most thoroughly tested, in the only way in which it can be practiced in the ordinary hives, this last plan of artificial swarming, and do not hesitate to say that it does not possess the very slightest practical value; and as this is the method which Apiarians have usually tried, it is not strange that they have almost unanimously pronounced Artificial swarming to be utterly worthless. The experience of Dzierzon on this point has been the same with my own.

Another method of artificial swarming has been zealously advocated, which, if it could only be made to answer, would be, of all conceivable plans the most effectual, and as it would require the smallest amount of labor, experience, or skill, would be everywhere practiced. A number of hives must be put in connection with each other, so as to communicate by holes which allow the bees to travel from any one apartment to the others. The bees, on this plan, are to *colonize themselves*, and in time, a single swarm will, of its own accord, multiply so as to form a large number of independent families, each one possessing its own queen, and all living in perfect harmony.

This method so beautiful and fascinating in theory, has been repeatedly tried with various ingenious modifications, but in every instance, as far as I know, it has proved an entire failure. It will always be found if bees are allowed to pass from one hive to another, that they will still, for the most part, confine their breeding operations to a single apartment, if it is of the ordinary size, while the others will be used, chiefly for the storing of honey. This is almost invariably the case, if the additional room is given by collateral or side boxes, as the queen seldom enters such apartments for the purpose of breeding. If the new hive is directly *below* that in which the swarm is first lodged, then if the connections are suitable, the queen will be almost certain to descend and lay her eggs in the new combs, as soon as they are commenced by the bees; in this case, the upper hive is almost entirely abandoned by her, and the bees store the cells with honey, as fast as the brood is hatched, as their instinct impels them always, if they can, to keep their stores of honey *above* the breeding cells. So long as bees have an abundance of room below their main hive, they will never swarm, but will use it in the way that I have described; if the room is on the sides of their hive, and very accessible, they seldom swarm, but if it is above them, they frequently prefer to swarm rather than to take possession of it. But in none of these cases, do they ever, *if left to themselves*, form separate and independent colonies.

I am aware that the Apiarian, by separating from the main hive with a slide, an apartment that contains brood, and directing to it by some artificial contrivance a considerable number of bees, may succeed in rearing an artificial colony; but unless all his hives admit of the most thorough inspection, as he can never know their exact condition, he must always work in the dark, and will be much more likely to fail than succeed. Success indeed can only be possible when a skillful Apiarian devotes a large portion of his time to watching and managing his bees, so as to *compel* them to colonize, and even then it will be very uncertain; so that this plausible theory to be reduced to even a most precarious practice, requires more skill, care, labor and time, than are necessary to manage the ordinary swarming hives.

The failure of so many attempts to increase colonies by artificial means, as well in the hands of scientific and experienced Apiarians, as under the direction of those who are almost totally ignorant of the physiology of the bee, has led many to prefer to use non-swarming hives. In this way, very large harvests of honey are often obtained from a powerful stock of bees; but it is very evident that if the increase of new colonies were entirely discouraged, the insect would soon be exterminated. To prevent this, the advocates of the non-swarming plan, must either have their bees swarm, to some extent, or rely upon those who do.

My hive may be used as a non-swarmer, and may be made more effectually to prevent swarming, than any with which I am acquainted: as in the Spring, (See No. 34. p. 104,) ample accommodations may be given to the bees, below their main works, and when this is seasonably done, swarming will *never* take place.

There are certain objections however, which must always prevent the non-swarming plan from being the most successful mode of managing bees. To say nothing of the loss to the bee-keeper, who has, after some years, only one stock, when if the natural mode of increase had been allowed, he ought to have a number, it is usually found that after bees have been kept in a non-swarming hive for several seasons, they seem to work with much less vigor than usual. Of this, any one may convince himself, who will compare the industrious working of a new swarm, with that of a much more powerful stock in a non-swarming hive. The former will work with such astonishing zeal, that to one unacquainted with the facts, it would be taken to be by far the more powerful stock.

As the fertility of the queen decreases by age, the disadvantage of using non-swarming hives of the ordinary construction, will be obvious. This objection to the system can be remedied in my hive, as the old queen can be easily caught and removed; but when hives are used in which this cannot be done, the Apiary, instead of containing a race of young queens in the full vigor of their reproductive powers, will contain many that have passed their prime, and these old queens may die when there are no eggs in the hive to enable the bees to replace them, and thus the whole colony will perish.

If the bee-keeper wishes to winter only a certain number of stocks, I will, in another place, show him a way in which this can be done, so as to obtain more honey from them, than from an equal number kept on the non-swarming plan, while at the same time, they may all be maintained in a state of the highest health and vigor.

I shall now describe a method of artificial swarming, which may be successfully practiced with almost any hive, by those who have sufficient experience in the management of bees.

About the time that natural swarming may be expected, a populous hive, rich in stores is selected, and what I shall call a *forced swarm* is obtained from it, by the following process. Choose that part of a pleasant day, say from 10 A. M. to 2 P. M., when the largest number of bees are abroad in the fields; if any bees are clustered in front of the hive, or on the bottom-board, puff among them a few whiffs of smoke from burning rags or paper, so as to force them to go up among the combs. This can be done with greater ease, if the hive is elevated, by small wedges, about one quarter of an inch above the bottom-board. Have an empty hive or box in readiness, the diameter of which is as nearly as possible, the same with that of the hive from which you intend to drive the swarm. Lift the hive very gently, and without the slightest jar, from its bottom-board; invert it and carry it in the same careful manner, about a rod from its old stand, as bees are always much more inclined to be peaceable, when removed a short distance, than when any operation is performed on the familiar spot. If the hive is carefully placed on the ground, upside down, scarcely a single bee will fly out, and there will be little danger of being stung. Timid and inexperienced Apiarians will, of course, protect themselves with a bee-dress, and they may have an assistant to sprinkle the hive gently with sugar-water, as soon as it is inverted. After placing the hive in an inverted position on the ground, the empty hive must be put over it, and every crack from which a bee might escape, must be carefully closed with paper or any convenient material. The upper hive ought to be furnished with two or three slats, about an inch and a half wide, and fastened one third of the distance from the top, so as to give the bees every opportunity to cluster.

As soon as the Apiarian is perfectly sure that the bees cannot escape, he should place an empty hive upon the stand from which they were removed, so that the multitudes which return from the fields may enter it, instead of dispersing to other hives, where some of them may meet with a very unkind reception; although as a general rule, a bee with a load of freshly gathered honey, after the extent of his resources is ascertained, is almost always, welcomed by any hive to which he may carry his treasures; while a poor unfortunate that ventures to present itself empty and poverty stricken, is generally at once destroyed! The one meets with as friendly a reception as a wealthy gentleman who proposes to take up his abode in a country village, while the other is as much an object of dislike as a pauper who is suspected of wishing to become a parish charge!

To return to our imprisoned bees. Beginning at the top, or what is now, (as the hive is upside down,) the bottom, their hive should be beaten smartly with two small rods on the front and back, or on the sides to which the combs are attached, so as to run no risk of loosening them. If the hive when removed from its stand was put upon a stool or table, or something not so solid as the ground, the drumming will cause more motion, and yet be less apt to start any of the combs. These "rappings" which certainly are not of a very "spiritual" character, produce nevertheless, a most decided effect upon the bees: their first impulse is to sally out and wreak their vengeance upon those who have thus rudely assailed their honied dome; but as soon as they find that they are shut in, a sudden fear that they are to be driven from their treasures, seems to take possession of them. If the two hives have glass windows, so that all the operations can be witnessed, the bees, in a few moments, will be seen most busily engaged in gorging themselves with honey. During all this time, the rapping must be continued, and in about five minutes, nearly every bee will have filled itself to its utmost capacity, and they are now prepared for their forced emigration; a prodigious hum is heard, and the bees begin to mount into the upper box. In about ten minutes from the time the rapping began, the mass of the bees with their queen will have ascended, and will hang clustered, just like a natural swarm. The box with the expelled bees must now be gently lifted off, and should be placed upon a bottom-board with a gauze wire ventilator, so that the bees may be confined, and yet have plenty of air. A shallow vessel or a piece of old comb containing water, ought to be first placed on the bottom-board. If no gauze wire bottom-board is at hand, the hive must be wedged up, so as to admit an abundance of air, and be set in a shady place.

The hive from which the bees were driven, must now be set, without crushing any of the bees, on its old spot, in the place of the decoy hive, that all the bees which have returned from abroad, may enter. Before this change is made, these bees will be running in and out of the empty hive, but as soon as the opportunity is given them, they will crowd into their well-known home, and if there are no royal cells started, will proceed, almost at once, to construct them, and the next day they will act as though the forced swarm had left of its own accord. When the operation is delayed until about the season for natural swarming, the hive will contain immature queens, if the bees were intending to swarm, and a new queen will soon take the place of the old one, just as in natural swarming. If it is performed too early, and before the drones have made their appearance, the young queen may not be seasonably impregnated, and the parent stock will perish.

It will be obvious that this whole process, in order to be successfully performed, requires a knowledge of the most important points in the economy of the bee-hive; indeed the same remark may be made of almost any operation, and those who are willing to remain ignorant of the laws which regulate the breeding of bees, ought not to depart in the least, from the old-fashioned mode of management. All such deviations will only be attended with a wanton sacrifice of bees. A man may use the common swarming hives a whole life-time, and yet remain ignorant of the very first principles in the physiology of the bee, unless he gains his information from other sources; while, by the use of my hives, any intelligent cultivator may, in a single season, verify for himself, the discoveries which have only been made by the accumulated toil of many observers, for more than two thousand years. The ease with which Apiarians may now, by the sight of their own eyes, gain a knowledge of all the important facts in the economy of the hive, will stimulate them most powerfully, to study the nature of the bee and thus to prepare themselves for an enlightened system of management.

In giving directions for the creation of forced swarms, I advised that it should be done during the pleasantest part of the day, when the largest number of bees are foraging abroad. If the operation is performed when all the bees are at home, and they are all driven into the empty hive, the old hive will be so depopulated that many of the young will perish for want of suitable attention, and the parent stock will be greatly deteriorated in value. If only a part of the bees are expelled, the queen may be left behind, and the whole operation will be a failure, and at best it will be difficult to make a suitable division of the bees between the two hives. Indeed, under any circumstances, this is the most difficult part of the process, and it often requires no little judgment to equalize the two colonies.

Some recommend placing the forced swarm on the old stand, and removing the parent hive with the bees that are deemed sufficient, to a new place. If this is done, and the bees have their liberty, so many of them will leave for the familiar spot, that the hive will be almost deserted, and a very large proportion of its brood will perish. The bees in this hive, if it is to be set in a new place, must have water given to them, and be so shut up as to have an abundance of air, until late in the afternoon of the third day, when the hive may be opened, and they will take wing, almost as though they were intending to swarm. Some will even then, return to the place where they originally stood, and join the forced swarm, but the most of them, after hovering in the air for a short time, will re-enter the hive. During the time that they have been shut up, thousands of young bees will have emerged from their cells, and these, knowing no other home, will aid in taking care of the larvæ, and in carrying on the work of the hive.

Instead of trying to make an equitable division at the time of driving out the bees, I prefer to expel all that I can, and to rely upon the bees returning from their gatherings, to replenish the old stock. If the number appears to be too small, I open temporarily the entrance of the hive containing the forced swarm, and permit as many as I judge best, to come out and enter their old abode. It must here be borne in mind, that bees which are thus ejected from a hive, do not, in all respects, act like a natural swarm, which having left the parent stock, of its own accord, never seeks, unless it has lost its queen, to return; whereas, many of the forced swarm, as soon as they leave the hive into which they have been driven, will return to their former abode. The same is true of bees which are moved to any distance not far enough to be beyond the limits of their previous excursions in search of food. If we could only make our bees when moved, or forced to swarm, adhere to their hives as faithfully as a natural swarm, many difficulties which now perplex us, would be at once removed.

Having ascertained that the parent hive contains a sufficient number of bees to carry on operations, about sun-set, after the bees are all at home, it may be removed to a new stand, and the bees, after being supplied with water, must be shut up, according to the directions previously given. If the hive is so constructed that water cannot be conveniently given them, the following plan I have found to answer most admirably. Bore a small hole towards the top on the front side, and with a straw, water may be injected with scarcely any trouble. A mouthful once or twice a day, will be sufficient. If the bees are confined without water, they will not be able to prepare the food for the larvæ, and multitudes of them must necessarily perish.

The expelled colony must be placed, on the same evening, precisely where the hive from which they were driven stood, and have their liberty given to them. The next morning, they will work with as much vigor as though they had swarmed in the natural way.

The directions which have here been given for creating forced swarms, will be found to differ in some important respects from any which other Apiarians have previously furnished. I have already shown that it is difficult to secure the right number of bees for the parent stock, unless it is set temporarily on its old stand, so as to catch up the returning bees. The common plan has been to try to leave in it, as many bees as are needed, and then to shut it up for a few days, having placed it in a new spot, while the forced swarm is immediately replaced so that all the stragglers may be added to it. If we could always be sure of driving out the queen, and with her, as many bees as we want and *no more*, this would undoubtedly be the simplest plan; but for the reasons already assigned, it will be found a very precarious operation.

Some Apiarians recommend putting the forced swarm in a new place in the Apiary; but as large numbers of the bees will be sure, when they go out to work, to return to the familiar spot, the new colony will often be so seriously depopulated as to be of but little value. If the Apiarian can remove his forced swarms, some two or three miles off, he may give them their liberty at once, and in the course of a few weeks, he can, without risk, bring them back to his Apiary.

If he chooses, he may allow the parent stock to remain on the old stand, and confine the forced swarm, until about an hour before sun set of the third day. They must in the mean time be supplied with both honey and water, and if they cannot be kept cool and quiet, they should be removed into the cellar until they are placed in their new position. Many will even then return to the old spot, but not enough to interfere seriously with their prosperity. If the bees cannot, as in my hives, be kept cool and dark, they will be excessively uneasy, and may suffer very seriously from so long confinement: hence the very great importance of setting them in the cellar.

It may seem strange, that bees, when their hive is moved, or when they are forcibly expelled from it, should not adhere to the new spot, just as when they have swarmed of their own accord. In each case, as soon as a bee leaves its new place, it flies with its head turned towards the hive, in order to mark the surrounding objects, that it may be able to return to the same spot; but when they have not emigrated of their own accord, many of them seem, when they rise in the air, or return from work, entirely to forget that their location has been changed; and they return to the place where they have lived so long, and if no hive is there, they often die on the deserted and desolate, yet home-like spot. If, on the contrary, they swarmed of their own accord, they seldom, if ever, make such a mistake. It may truly be said that

"A 'bee removed' against its will Is of the same opinion still."

I have been thus minute in describing the whole process of creating forced swarms, not merely on account of the importance of the plan in multiplying colonies, but because the driving or drumming out of bees from a common hive, is employed with great success in a variety of ways which will be hereafter specified. I doubt not that many bee-keepers, on reading this mode of creating colonies, are ready to object that it not only requires more skill, but more time and labor, than to allow them to swarm, and then to hive them in the old-fashioned way.

As practiced with ordinary hives, it is undoubtedly liable to this serious objection, and I would easily with my basket hiver, undertake to hive four natural swarms, in the time that it would require to create one forced swarm; to say nothing of the care which must be bestowed upon the artificial swarms, with their parent stocks, after the driving process has been completed. For this reason, I do not advise the bee-keeper to force his swarms from the common hives, until he has first ascertained that they are not likely to swarm in tolerably good season, of their own accord, unless he is afraid that they will come out during his absence, and decamp to the woods.

By the aid of my hives, this process may be most expeditiously performed. An empty hive, with its frames furnished with guide combs, must be in readiness. The cover of the full hive should be removed, and the bees gently sprinkled with sugar-water from a watering pot that discharges a fine stream. In about two minutes, the frames may be taken out, and the bees, by a quick motion, shaken on a sheet directly in front of their hive. As fast as a comb is deprived of its bees, it should be set in a proper position in the new hive, and an empty frame put in its place. Two or three of the combs containing brood, eggs, &c., should be left in the old hive, as well to give them greater encouragement, as to prevent them from being dissatisfied if their queen should, by any possibility, be taken from them. In removing the frames with the bees, I always look for the queen, and if I see her, as I generally do, I return to the hive the frame which contains her, without shaking off the bees. In that case, I put several of the necessary combs into the new hive, with all the bees upon them.

In dislodging the bees upon the sheet, I do not shake them all off from the frames; but leave about one quarter of them on, and put them with the combs into the new hive. I never knew the queen to be left on a frame after it was shaken so that the larger portion of the bees would fall off. As soon as the operation is completed, and the necessary number of bees have been transferred with their comb to the new hive, it should be managed according to the directions previously given, in the case of the old hive from which a swarm was drummed out.

If in the operation the Apiarian does not see the queen, he must, in the course of the third day, examine the hive having the larger portion of bees, and if they have commenced building royal cells among the combs given to them, he may be certain that she is in the other hive. The comb containing the royal cells may then be transferred to that hive, and the queen searched for, and returned with the combs on which she is found, to her proper place. A little experience, however, will enable the operator to be sure from the first, that the queen is with the right division.

To most persons, it would seem to be of little consequence, in which hive the queen is placed: but if the bees which have only a few frames of comb, are compelled to rear another, they will be sure to fill their hive with comb unfit for breeding purposes, and will also be so long before they can have additions to their number, as to be of but little value.

If many swarms are to be created in this manner, and the operation is delayed until near swarming time, in some of them, numerous royal cells will be found, so that each stock which has no queen, may have one nearly mature, given to it, and thus much valuable time may be saved.

By making a few forced swarms, about a week or ten days before the time in which the most will be made, the Apiarian may be sure of having an abundance of sealed queens almost mature, so that every swarm may have one. If he can give each hive that needs it, an unhatched queen, without removing her from her frame, so much the better; but if he has not enough frames with sealed queens, while some of them contain two or more queens, he must proceed as follows:

With a very sharp knife, carefully cut out a queen cell, on a piece of comb an inch or more square; cut a place in one of the combs of the hive to which this cell is to be given, just about large enough to receive it in a natural position, and if it is not secure, put a little melted wax with a feather, where the edges meet. The bees will soon fasten it, so as to make all right. Unless very great care is used in transferring these royal cells, the enclosed queens will be destroyed; as their bodies, until they are nearly mature, are so exceedingly soft, that a very slight compression of their cell often kills them. For this reason, I prefer not to remove them, until they are within three or four days of hatching. As the forcing of a swarm may always be conducted, with my hives, in such a manner that the Apiarian can be sure to effect a suitable division of the bees, the process may be performed at any time when the sun is above the horizon, and the weather is not too unpleasant. It ought not to be attempted when the weather is so cool as to endanger the destruction of the brood, by a chill; and never unless when there is not only sufficient light to enable the Apiarian to see distinctly, but enough for the bees that take wing, to see the hive, and direct their flight to its entrance. If hives are meddled with, when it is dark, the bees are always more irascible, and as they cannot see where to fly, they will constantly be alighting upon the person of the bee-keeper, who will be almost sure to receive some stings. I have seldom attempted night-work upon my bees, without having occasion most thoroughly to rue my folly. If the weather is not too cool, early in the morning, before the bees are stirring, will be the best time, as there will be less danger of annoyance from robber-bees.

If honey-water is used instead of sugar-water in sprinkling the bees when the hive is first opened, the smell will be almost certain to entice marauders from other hives to attempt to take possession of treasures which do not belong to them, and when they once commence such a pilfering course of life, they will be very loth to lay it aside. When the honey harvest is abundant, (and this is the very time for forcing swarms,) bees, with proper precautions, are seldom inclined to rob. I have sometimes found it difficult to induce them to notice honey-combs which I wished them to empty, even when they were placed in an exposed situation. This subject, however, will be more fully treated in the remarks on Robbing.

Perhaps some of my readers will hardly be able to convince themselves that bees may be dealt with after the fashion I have been describing, without becoming greatly enraged; so far is this from being the case, that in my operations, I often use neither sugar-water nor bee-dress, although I do not recommend the neglect of such precautions.

The artificial swarm may be created with perfect safety, even at mid-day, when thousands of bees are returning to the hive: for these bees being laden with honey, never venture upon making an attack, while those at home may be easily pacified.

I find a very great advantage in the peculiar shape of my hive, which allows the top to be easily removed, and the sugar-water to be sprinkled upon the bees, before they attempt to take wing. If like the Dzierzon hive, it opened on the end, it would be impossible for me to use the sweetened water, so as to make it run down between all the ranges of comb, and I should be forced, as he does, to employ smoke, in all my operations. Huber thus speaks of the pacific effect produced upon the bees by the use of his leaf hive. "On opening the hive, no stings are to be dreaded, for one of the most singular and valuable properties attending my construction, is its rendering the bees tractable. I ascribe their tranquility to the manner in which they are affected by the sudden admission of light, they appear rather to testify fear than anger. Many retire, and entering the cells, seem to conceal themselves." I will admit that Huber has here fallen into an error which he would not have made, had he used his own eyes. The bees do indeed enter the cells when the frames are exposed, but not "to conceal themselves;" they imagine that their sweets, thus unceremoniously exposed to the light of day, are to be taken from them, and they fill themselves to their utmost capacity, in order to save all that they can. I always expect them to appropriate the contents of the open cells, as soon as I remove their frames from the hive. It is not merely the *sudden* admission of light, but its introduction from an *unexpected quarter*, that seems for the time to disarm the hostility of the bees. They appear for a few moments, almost as much confounded as we should be, if without any warning the roof and ceiling of a house should suddenly fly off into the air. Before they recover from their amazement, the sweet libation is poured out upon them, and surprize is quickly converted into pleasure rather than anger. I believe that in the working season, almost all the bees near the top are gorged with honey, and that this is the reason why opening the hive from ABOVE is so easily effected. The bees below that are disposed to resent any intrusions, are met in their threatening ascent, with an avalanche of nectar which "like a soft answer," most effectually "turneth away wrath." Who would ever be willing to use the sickening fumes of the disgusting weed, when so much pleasure instead of pain may be given to his bees. That bees never seem to be prepared to make an instant assault from the top of their hive, but only near the entrance, any one may be convinced of, who will put my frames into a suspended hive with a movable bottom which may be made to drop at pleasure. If now, for any purpose, he attempts to meddle with the combs from below, he will find that unless he uses smoke, the bees will be almost, if not quite unmanageable.

I shall now give some directions, which will greatly assist the Apiarian in his operations. He must bear in mind that nothing irritates bees more than a sudden jar, and that this must, in all cases, be most carefully avoided. The inside cover of the hive, or as I shall term it, the *honey-board*, because the surplus honey receptacles stand upon it, can never be very firmly attached by the bees: it may always be readily loosened with a thin knife, or better still, with an apothecary's spatula, which will be very useful for many purposes in the Apiary. When the honey-board is removed, its lower surface will be usually covered with bees, and it should be carefully set on end, so as not to crush them. There is not the least danger that one of them will offer to sting, as they are completely bewildered by the sudden introduction of light, and their removal from the hive. As soon as the cover is disposed of, the Apiarian should sprinkle the bees with the sweet solution. This should descend from the watering-pot in a fine stream, so as not to *drench* the bees, and should fall upon the tops of the frames, as well as between the ranges of comb. The bees will at once, accept the proffered treat, and will begin lapping it up, as peaceably as so many chickens helping themselves to corn. While they are thus engaged, the frames must be very gently pried by a stick, from their attachments to the rabbets on which they rest; this may be done without any jar and without wounding or enraging a single bee. They may all be loosened preparatory to removing them, in less than a minute.[17] By this time, the sprinkled bees will have filled themselves, or if all have not done so, the grateful intelligence that sweets have been furnished them, will diffuse an unusual good nature through all the honied realm. The Apiarian should now remove one of the outside frames, taking hold of its two ends which rest upon the rabbets, and carefully lifting it out without inclining it from its perpendicular position, so as not to injure a single bee. The removal of the next comb, and of all the succeeding ones, will be more easily effected, as there will be more room to operate to advantage. If bees were disposed to fly away at once from their combs, as soon as they were taken out, it would be very difficult to manage them, but so far are they from doing this, that they adhere to them with most wonderful tenacity. I have sometimes removed all the combs, and arranged them in a continued line, and the bees have not only refused to leave them, but have stoutly defended them against the thieving propensities of other bees. By shaking off the bees from the combs upon a sheet, and securing the queen, I can, on any pleasant day, exhibit nearly all the appearances of natural swarming. The bees, as soon as they miss their queen, will rise into the air, and by placing her on the twig of a tree, they will soon cluster around her in the manner already described.

A word as to the manner of catching the queen. I seize her very gently, as I espy her among the bees, and by taking care to crush none of them, run not the least risk of being stung. The queen herself never stings, even if handled ever so roughly.

In removing the frames from the hive, it will be found very convenient to have a box with suitable rabbets in which they may be temporarily put, and covered over with a piece of cotton cloth. They may thus be very easily protected from the cold, and from robbing bees, if they are to be kept out of the hive for some time; and such a box will be very convenient to receive frames that are lifted out for examination. In returning the frames to a hive, care must be taken not to crush the bees where their ends rest upon the rabbets; they must be put in slowly, so that a bee, when he feels the slightest pressure may have a chance to creep from under them, before he is hurt.

The honey-cover, for convenience, is generally in two pieces: these cannot be laid down on the hive, without danger of killing many bees; they are therefore very carefully *slid* on, so that any bees which may be in the way, are pushed before them, instead of being crushed. If any bees are upon such parts of the hive as to be imprisoned if the outside cover is closed, it should be left a little open, until they have flown to the entrance of the hive. It cannot be too deeply impressed upon the bee-keeper, that all his motions must be slow and gentle, and that the bees must not be injured or breathed upon. If he will carefully follow the directions I have given, he may soon open a hundred hives and perform any necessary operation upon them, without any bee-dress, and yet with very little risk of being stung, but I almost despair of being able to convince even the most experienced Apiarians, of the ease and safety with which bees may be managed on my plan, until they have actually been eye-witnesses of its successful operation.

I can make an artificial colony in the way above described in ten minutes from the time that I open the hive, and if I see the queen as quickly as I often do, in not more than five minutes. Fifteen minutes will be a very liberal allowance of time to complete the whole work. If I had an Apiary of a hundred colonies, in less than a week, if the weather was pleasant, I could without any assistance easily finish the business of swarming for the whole season.

But how can the Apiarian, if he delays the formation of artificial swarms until nearly the season for natural swarming, be sure that his bees will not swarm in the usual way? Must he not still be constantly on hand, or run the risk of losing many of his best swarms? I come now to the entirely novel plan by which such objections are completely obviated. If the Apiarian decides that he can most advantageously multiply his colonies by artificial swarming, he must see that all his fertile queens are deprived of their wings, so as to be unable to lead off new swarms. As an old queen never leaves the hive except to accompany a new swarm, the loss of her wings does not, in the least interfere with her usefulness, or with the attachment of the bees. Occasionally, a wingless queen is so bent on emigrating, that in spite of her inability to fly, she tries to go off with a swarm; she has "a will," but contrary to the old maxim she can find "no way," but helplessly falls upon the ground instead of gaily mounting into the air. If the bees succeed in finding her, they will never desert her, but cluster directly around her, and may thus be easily secured by the Apiarian. If she is not found, the bees will return to the parent stock to await the maturity of the young queens. The Apiarian will ordinarily be prepared to form his artificial colonies before any of these young queens are hatched.

The following is the best plan for removing the wings from the queens. Every hive which contains a young queen, ought to be examined about a week after she has hatched, (see Chapter on Loss of the Queen,) in order to ascertain that she has been impregnated, and has begun to lay eggs. Some of the central combs or those on which the bees are most thickly clustered, should be first lifted out, for she will almost always be found on one of them; the Apiarian when he has caught her, should remove the wings on one side with a pair of scissors taking care not to hurt her. On examining his hives next season, let him remove one of the two remaining wings from the queen. The third season, he may deprive her of her last wing. Bees always have four wings, a pair on each side. This plan saves him the trouble of marking his hives so as to know the age of the queens they contain.

As the fertility of the queen generally decreases after the second year, I prefer, just before the drones are destroyed, to kill all the old queens that have entered their third year. In this way, I guard against some of my stocks becoming queenless, in consequence of the queen dying of old age, when there is no worker-brood in the hive, from which they can rear another: or of having a worthless, drone-laying queen whose impregnation has been retarded. These old queens are removed at that period of the year when their colony is strong in numbers; and as the honey-harvest is by this time, nearly over, their removal is often a positive benefit, instead of a loss. The population is prevented from being over crowded at a time when the bees are consumers and not producers, and when the young queen, reared in the place of the old one matures, she will rapidly fill the cells with eggs, and raise a large number of bees to take advantage of the late honey-harvest, and to prepare the hive to winter most advantageously.

The certainty, rapidity and ease of making artificial swarms with my hives, will be such as to amaze those most who have had the greatest experience and success in the management of bees. Instead of weeks wasted in watching the Apiary, in addition to all the other vexations and embarrassments which are so often found to attend reliance on natural swarming, the Apiarian will find not only that he can create all his new colonies in a very short time, but that he can, if he chooses, entirely prevent the issue of all after-swarms. In order to do this, he ought to examine the stocks which are raising young queens, in season to cut out all the queen cells but one, before the larvæ come to maturity. If he gave them a sealed queen nearly mature, they will raise no others, and no swarming, for that season, will take place. If the Apiarian wishes to do more than to double his stocks in one season, and is favorably situated for practicing natural swarming, he can allow the stocks that raise young queens to swarm if they will, and he can strengthen the small swarms by giving to them comb with honey and maturing brood from other hives. Or he can, after an interval of about three weeks, make one swarm from every two good ones in his Apiary, in a way that will soon be described.

I do not know that I can find a better place in which to impress certain highly important principles upon the attention of the bee-keeper. I am afraid, that in spite of all that I can say, many persons as soon as they find themselves able to multiply colonies at pleasure, will so overdo the matter, as to run the risk of losing all their bees. If the Apiarian aims at obtaining a large quantity of honey in any one season, he cannot at the furthest, more than double the number of his stocks: nor can he do this, unless they are all strong, and the season favorable. The moment that he aims, in any one season, at a more rapid increase, he must not only renounce the idea of having any surplus honey, but must expect to purchase food for the support of his colonies, unless he is willing to see them all perish by starvation. The time, food, care and skill required to multiply stocks with very great rapidity, in our short and uncertain climate, are so great that not one Apiarian in a hundred can expect to make it profitable; while the great mass of those who attempt it, will be almost sure, at the close of the season, to find themselves in possession of stocks which have been so managed as to be of very little value.

Before explaining some other methods of artificial swarming, which I have employed to great advantage, I shall endeavor to impress upon the mind of the bee-keeper, the great importance of thoroughly understanding each season, the precise object at which he is aiming, before he enters on the work of increasing his colonies. If his object is, in any one season, to get the largest yield of surplus honey, he must at once make up his mind to be content with a very moderate increase of stocks. If, on the contrary, he desires to multiply his colonies, say, three or four fold, he must be prepared, not only to relinquish the expectation of obtaining any surplus honey, if the season should prove unfavorable, but to purchase food for the support of his bees. Rapid multiplication of colonies, and large harvests of surplus honey cannot, in the very nature of things, be secure in our climate, in any one season.

If the number of colonies is to be increased to a large extent, then the bees in the Apiary will be tasked to the utmost in building new comb, as well as in rearing brood. For these purposes, they must consume the supply of honey which, under other circumstances, they would have stored up, a part for their own use in the main hive, and the balance for their owner, in the spare honey-boxes.

To make this matter perfectly plain, let us suppose a colony to swarm. If the new hive, into which the swarm is put, holds, as it ought, about a bushel, it will require about two pounds of wax to fill it with comb, and at least forty pounds of honey will be used in its manufacture! If the season is favorable, and the swarm was large and early, they may gather, not only enough to build this comb and to store it with honey sufficient for their own use, but a number of pounds in addition, for the benefit of their owner. If the old stock does not swarm again, it will rapidly replenish its numbers, and as it has no new comb to build in the main hive which already contains much honey, it will be able to store up a generous allowance in the upper boxes. These favorable results are all on the supposition that the season was ordinarily productive in honey, and that the hive was so powerful in numbers as to be able to swarm early. If the season should prove to be very unfavorable, the first swarm cannot be expected to gather more than enough for its own use, while the parent stock will yield only a small return. The profits of the bee-keeper, in such an unfortunate season, will be mainly in the increase of his stocks. If the swarm was late, in consequence of the stock being weak in Spring, the early part of the honey-harvest will pass away, and the bees will be able to obtain from it, but a small share of honey. During all this time of comparative inactivity, the orchards may present

"One boundless blush, one white empurpled shower Of mingled blossoms," and tens of thousands of bees from stronger stocks, may be engaged all day, in sipping the fragrant sweets, so that every gale which "fans its odoriferous wings" about their dwellings, dispenses

"Native perfumes, and whispers whence they stole[18] Those balmy spoils."

By the time that the feeble stock is prepared to swarm, if it swarm at all that season, the honey-harvest is almost over, and the new colony will seldom be able to gather enough for its own use, so that unless fed, it must perish the succeeding Winter. Bee-keeping with colonies feeble in the Spring, is most emphatically nothing but "folly and vexation of spirit."

I have shown how the bee-keeper, with a strong stock-hive which has swarmed early and but once, may in a favorable season realize handsome profits from his bees. If the parent stock throws a second swarm, then, as a general rule, unless this swarm was very early, and the honey season good, if managed on the ordinary plan, it will seldom prove of any value. It will almost always perish in the Winter, if it does not desert its hive in the Fall, and the family from which it issued, will not only gather no surplus honey, (unless it was secured before the first swarm issued,) but will very often perish likewise. Thus the inexperienced owner who was so delighted with the rapid increase of his colonies, begins the next season with no more colonies than he had the year before, and has very often lost all the time he has bestowed upon his bees. I can, to be sure, on my plan, prevent the death of the bees, and can build up all the feeble colonies, so as to make them strong and powerful: but only by giving up all idea of obtaining a single pound of honey. From the first swarm, I must take combs containing maturing brood, to strengthen my weak swarms, and this first swarm however powerful or early, instead of being able to store its combs with honey, will be constantly tasked in building new combs to replace those taken away, so that when the honey harvest closes, it will have scarcely any honey, and must be fed to prevent it from starving. Any man who has sense enough to be entrusted with bees, can, from these remarks, understand exactly why it is impossible to multiply colonies rapidly in any one season, and yet obtain from them large supplies of honey. Even the doubling of stocks in one season, will very often be too rapid an increase, if the greatest quantity of spare honey is to be obtained from them; and when the largest yield of honey is desired, I much prefer to form, in a way soon to be described, only one new stock from two old ones; this will give even more from the three, than could have been obtained from the two, on the ordinary non-swarming plan.

I would very strongly dissuade any but experienced Apiarians, from attempting at the furthest, to do more than to triple their stocks in one year. In order to furnish directions for very rapid multiplication, sufficiently full and explicit to be of any value to the inexperienced, I should have to write a book on this one topic; and even then, the most of those who should undertake it, would be sure at first to fail.

I have no doubt that with ten strong stocks of bees in a good location, in one favorable season, I could so increase them as to have, on the approach of Winter, one hundred good colonies: but I should expect to feed hundreds of pounds of honey, to devote nearly all my time to their management, and to bring to the work, the experience of many years, and the wisdom acquired by numerous failures. After all, what we most need, in order to be successful in the cultivation of bees, is a *certain*, rather than a *rapid* multiplication of stocks. It would require but a very few years to stock our whole country with bees, if colonies could only be doubled annually; and an increase of even one third, would before long, give us bees enough. This rate of increase I should always encourage in the swarming season, even if, in the Fall, I reduced my stocks (see Union of Stocks) to the Spring number. In the long run, it will keep the colonies in a much more prosperous condition, and secure from them the largest yield of honey.

I have never myself hesitated to sacrifice one or more colonies, in order to ascertain a single fact, and it would require a separate volume quite as large as this, to detail the various experiments which I have made on the subject of Artificial Swarming. The practical bee-keeper, however, should never, for a moment, lose sight of the important distinction between an Apiary managed principally for the purposes of experiment and discovery, and one conducted almost exclusively with reference to pecuniary profit. Any bee-keeper can easily experiment with my hives: but I would recommend him to do so, at first, on a small scale, and if profit is his object, to follow the directions furnished in this treatise, until he is *sure* that he has discovered others which are preferable. These cautions are given to prevent persons from incurring serious losses and disappointments, if they use hives which, if they are not on their guard, may tempt them into rash and unprofitable courses, by allowing so easily of all manner of experiments. Let the practical Apiarian remember that the less he disturbs the stocks on which he relies for surplus honey, the better. After they are properly lodged in their new hive, they ought by all means to be allowed to carry on their labors without any interruption. The object of giving the control over every comb in the hive, is not to enable him to be incessantly taking them in and out, and subjecting the bees to all sorts of annoyances. Unless he is conducting a course of experiments, such interference will be almost as silly as the conduct of children who pull up the seeds which they have planted, to see whether they have sprouted, or how much they have grown. If after these cautions, any still choose to disregard them, the blame of their losses will fall, not upon the hive, but upon their own mismanagement.

Let me not, for a moment, be understood as wishing to discourage investigation, or to intimate that perfection has been so nearly attained that no more important discoveries remain to be made. On the contrary, I should be glad to learn that many who have the time and means, are disposed to use the facilities furnished by hives which give the control of each comb, to experiment on a large scale; and I hope that every intelligent bee-keeper who follows my plans, will experiment at least on a small scale. In this way, we may soon expect to see, more satisfactorily elucidated, some points in the Natural History of the bee, which are still involved in doubt.

Having described the way in which forced swarms are made, both in common hives and in my own, when the Apiarian wishes in one season merely to double his colonies, I shall now show in what way he can secure the largest yield of honey, by forming only one new colony from two old ones.

Early in the season, before the bees fly out, or better still, after they ceased to fly in the previous Fall, the two hives from which the new colony is to be formed, should be placed near each other, unless they are already, not more than a foot apart. When the time for forming the artificial colony has arrived, these hives should be removed from their stand, and the bees driven from them, precisely in the manner already described. If all the bees are at home, I sometimes shut up the hives on their stand, and drum long enough to cause the bees to fill themselves before the hive is removed. Timid Apiarians may find some advantage in this course, as the bees will all be quiet after they are well drummed, and the hive may then be removed with greater safety. In five minutes I can in this way reduce any swarm to a peaceable condition. After the forced swarms are secured, the removed hives are replaced, in order to catch up all the returning bees, and the forced swarms must be shut up, until towards sunset; unless it is judged best to keep the entrances temporarily open, so as to secure the return of a sufficient number of bees to the parent stocks. The old stocks are now moved to a new place, and managed according to the previous directions. If neither of the expelled swarms was driven into the hive intended for the new colony, then the proper hive must be placed, as near as possible, in the center of the space previously occupied by the original colonies. One of the swarms must now be shaken out upon a sheet, in front of this hive which should be elevated, so as to enable the bees to enter it readily. As soon as they are shaken out, they should be gently sprinkled with sugar-water scented with peppermint, or any other fragrant odor. Diligent search must now be made for the Queen, and if found, she should be carefully removed, and given to the hive to which she belongs. If the queen of the first swarm has been found, the second colony may be shaken out, and sprinkled in the same way, and allowed to enter without any further trouble. If the queen of the first colony was not found, then that of the second one must be sought for; if neither can be found, (though this, after a little experience, will very seldom happen,) one of the Queens will soon kill the other, and reign over the united family. The next day, the doubled colony will be found working with amazing vigor, and it will not only fill its main hive, but will, in any ordinary season, gather large quantities of surplus honey besides.

The Apiarian who relies upon natural swarming, can double his new colonies if they issue at the same time, by hiving them together, or if this cannot be done, he may hive them in separate hives, and then, towards evening, set one hive on a sheet, and shake down the bees from the other, so that they can enter and join the first. It may be safely done, even if several days have elapsed before the second colony swarms; although in this case, I prefer after turning up their hive to sprinkle the oldest swarm with scented sugar-water, and then to give the new swarm the same treatment. I have doubled natural swarms in this way, repeatedly, and have never, when they were early, failed to secure from them a large quantity of honey. In sprinkling bees, let the operator remember that they are not to be *drenched*, or almost drowned, as in this case, they will require a long time to enter the hive. Bees seem to recognize each other by the sense of smell; and when they are made to have the same odor, they will always mingle peaceably. This is the reason why I use a few drops of peppermint in the sugar-water.

If one of the queens of the forced swarms can be returned to her own colony, it will of course, save them the time which would otherwise be lost in raising another. I do not know that I can better illustrate the importance of the inexperienced Apiarian following carefully my directions, than by supposing him to return the queen to the colony to which she does not belong. Now I can easily imagine that some bee-keeper may do so, conceiving that I am foolishly precise in my directions, and that the queen might be just as well given to one hive as to the other. But if this is done before at least 24 hours have elapsed since they were deprived of their own, she will almost certainly be destroyed. The bees do not *sting* a queen to death, but have a curious mode of crowding or knotting around her, so that she is soon smothered; and while thus imprisoned, she will often make the same piping note which has already been described. In all this treatise, I have constantly aimed to give no directions which are not important; and while I utterly repudiate the notion that these directions may not be modified and improved, I am quite certain that this cannot be done by any but those who have considerable experience in the management of bees.

The formation of one new swarm from two old colonies, may, of course, be very much simplified by the use of my hives. The two old hives are first opened and sprinkled, and the bees taken from them and put into the new hive in the same way in which the process was conducted when only one colony was expelled, some brood comb being given to the united family. There will be no difficulty in rightly proportioning the bees; one queen may always be caught and preserved, and the operation may be performed at any time when the sun is above the horizon. I have no doubt that those who have a strong stock of bees, and who are anxious to realize the largest profits in honey, will find this mode of increase, by far the simplest and best. If judiciously practiced, they will find that their colonies may always be kept powerful, and that they may be managed with very great economy in time and labor. As Apiarians may be so situated as to wish to increase their bees quite rapidly, I shall give such methods as from numerous experiments, many of them conducted on a large scale, I have found to be the best. I wish it however to be most distinctly understood, that I do not consider *very* rapid multiplication as likely to succeed, except in the hands of skillful Apiarians; and under ordinary circumstances it requires too much time, care and honey, to be of very great practical value. Its chief merit consists in the short time which it requires to build up an Apiary. After trying my mode of management for a few seasons, a bee-keeper may find out, that he is in all respects, favorably situated for taking care of a large stock of bees. Suppose him to have acquired both skill and confidence, and that he has ten powerful colonies. If he is willing to do without surplus honey for one season, and the honey-harvest should be very productive, he may without feeding, and without very much labor, safely increase his ten colonies to thirty. If he chooses to feed largely, he may *possibly* end the season with fifty or sixty, or even more; but he will *probably* end it in such a manner as most thoroughly to disgust him with his folly, and to teach him that in bee-keeping, as well as in other things, "Haste makes waste."

On the supposition that by the time the fruit-trees are in blossom, the Apiarian has, in hives of my construction, ten powerful colonies, let him select four of the strongest, and make from each a forced swarm. He will now have four queenless colonies, which will at once, proceed to supply themselves with a young queen. In about ten days, he may make from his other six stocks, six more forced swarms. He will probably find in making these, many sealed queens, if he has delayed the operation until about swarming time; so that he may give to each of the six stocks from which he has expelled a swarm, the means of soon obtaining another. If he has not enough for this purpose, he must take the required number from the four stocks which are raising young queens, the exact condition of which ought to have been previously ascertained. Some of these stocks will be found to contain a large number of queen cells. Huber, in one of his experiments, found twenty-four in one hive, and even a larger number has sometimes been reared by a single colony. As the Apiarian will always have many more queens than are wanted, he ought to select those combs which contain a sealed queen, so as to secure say, about fifteen combs, each of which has one or more queens. If necessary, he can cut out some of the cells, and adjust them in the manner previously described. Each comb containing a sealed queen must be put with all the bees adhering to it, into an empty hive; and by a divider, or movable partition, they must be confined to about one quarter of the hive; water should be given to them, and honey, if none is contained in the comb. I always prefer to select a comb which contains a large number of workers almost mature, and some of which are just beginning to hatch, so that even if a considerable number of the bees should return to the parent stock, after their liberty is given them, there will still be a sufficient number hatched, to attend to the young, and especially to watch over the maturing queens. If the comb contains a large number of bees just emerging from their cells, I prefer to confine them only one day, otherwise I keep them shut up until about an hour before sunset of the third day. The hives containing the small colonies, ought, if they are not well protected by being made double, to be set where they are thoroughly sheltered from the intense heat of the sun; and the ventilators should give them an abundance of air. They should also be closed in such a manner, as to keep the interior in entire darkness, so that the bees may not become too uneasy during their confinement. I accomplish this by shutting up their entrance, and replacing their front board, just as though I were intending to put them into winter quarters.

These small colonies I shall call *nuclei*, and the system of forming stocks from them, my nucleus system; and before I describe this system more particularly, I shall show other ways in which the nuclei can be formed. If the Apiarian chooses, he can take a frame containing bees just ready to mature, and eggs and young worms, all of the worker kind, together with the old bees which cluster on it, and shut them up in the manner previously described; even if he has no sealed queen to give them. If all things are favorable, they will set about raising a queen in a few hours. I once took not more than a tea-cup full of bees and confined them with a small piece of brood comb in a dark place, and found that in about an hour's time, they had begun to enlarge some of the cells, to raise a new queen! If the Apiarian has sealed queens on hand, they ought, by all means, to be given to the nuclei, in order to save all the time possible.

I sometimes make these nuclei as follows. The suitable comb with bees &c., is taken from a stock-hive, and set in an empty one, made to stand partly in the place of the old hive, which, of course, must previously be moved a little on one side. In this way, I am able to direct a considerable number of the bees from the old stock to my nucleus, and the necessity of shutting it up, is done away with. If the bees from the old stock do not enter the small one, in sufficient numbers, I sometimes close their hive, so that the returning bees can find no other place to enter. My object is not to catch up a *large* number of bees. For reasons previously assigned, I do not want enough to build new comb, but only enough to adhere to the removed comb, and raise a new queen from the brood, or develop the sealed one which has been given them. A short time after one nucleus has in this way, been formed, another may be made by moving the old hive again, and so a third or fourth, if so many are wanted. This plan requires considerable skill and experience, to secure the right number of bees, without getting too many.

If bees are to be made to enter a new hive, by removing the old one from its stand, it will always be very desirable not only to have the new one contain a piece of comb, but a considerable number of bees *clustered* on that comb. I repeatedly found my bees, after entering the hive, refuse to have anything to do with the brood comb, and for a long time, I was unable to conjecture the cause; until I ascertained that they were dissatisfied with its deserted appearance, and that, by taking the precaution to have it well covered with bees, I seldom failed to reconcile them to my system of forced colonization. I can usually tell, in less than two minutes, whether the operation will succeed or not. If the returning bees are content, they will, however much agitated at first, soon begin to join the cluster on the comb; while if they are dissatisfied, they will abandon the hive, and nearly all the bees that were originally on the comb, will leave with them. They seem capricious in this matter, and are sometimes so very self-willed, that they refuse to have anything to do with the brood comb, when I can see no good reason why they should be so rebellious.

I shall here state some *conjectures* which have occurred to me on this subject. Is it absolutely certain that bees can raise a queen from *any* egg or young larva which would produce a worker? Or if this is possible, is it certain that *any kind of workers* can accomplish this? Huber ascertained to his own satisfaction that there were two kinds of workers in a hive. He thus describes them.

"One of these is, in general, destined for the elaboration of wax, and its size is considerably enlarged when full of honey; the other immediately imparts what it has collected to its companions, its abdomen undergoes no sensible change, or it retains only the honey necessary for its own subsistence. The particular function of the bees of this kind is to take care of the young, for they are not charged with provisioning the hive. In opposition to the wax workers, we shall call them small bees or nurses."

"Although the external difference be inconsiderable, this is not an imaginary distinction. Anatomical observations prove that the capacity of the stomach is not the same—experiments have ascertained that one of the species cannot fulfil all the functions shared among the workers of a hive. We painted those of each class with different colors, in order to study their proceedings; and these were not interchanged. In another experiment, after supplying a hive deprived of a queen with brood and pollen, we saw the small bees quickly occupied in nutrition of the larvæ, while those of the wax working class neglected them. Small bees also produce wax, but in a very inferior quantity to what is elaborated by the real wax workers."

Now if these statements can be relied on, and thus far I have nearly always found Huber's statements, where-ever I had an opportunity to test them, to be most wonderfully reliable, then it may be that when bees refuse to cluster on the brood comb and to proceed at once to rear a new queen, it is because they find that some of the conditions necessary for success are wanting. Either there may not be a sufficient number of wax-workers, to enlarge the cells, or a sufficient number of nurses to take charge of the larvæ; or it may be that the cells contain only young wax-workers which cannot be developed into queens, or only young nurses, which may be in the same predicament.

If any of my readers imagine that the work of carefully experimenting, in order to establish facts upon the solid basis of complete demonstration, is an easy work, let them attempt now to prove or disprove the truth of any or all of my conjectures upon this single topic. They will probably find the task more difficult than to blot over whole quires and reams of paper with careless assertions.

All operations of any kind which interfere in the very least, with the natural mode of forming colonies, are best performed in the swarm-ing season: or at least, at a time when the bees are breeding freely, and are able to bring in large stores of honey from the fields. At other times, they are very precarious, and unless under the management of persons who have great experience, they will in most cases, end in nothing but vexatious losses and disappointments.

It is quite amusing to see how bees act, when they find, on their return from foraging abroad, that their hive has been moved, and another put in its place. If the new hive is precisely similar to their own, in size and outward appearance, they enter it as though all was right; but in a few moments, they rush out in violent agitation, imagining that they have made a prodigious mistake and have entered the wrong place. They now take wing again in order to correct their blunder, but find to their increasing surprise, that they had previously directed their flight to the familiar spot; again they enter, and again they tumble out, in bewildered crowds, until, at length, if they can find the means of raising a new queen, or one is already there, they seem to make up their minds that if this is not home, it not only looks like it, but stands just where their home ought to be, and is at all events the only home they are likely to get. No doubt they often feel that a very hard bargain has been imposed upon them, but they seem generally determined to make the best of it.

There is one trait in the character of bees, for which I feel, not merely admiration, but the most profound respect. Such is their indomi-table energy and perseverance, that under circumstances apparently the most despairing, they will still labor to the utmost, to retrieve their losses, and sustain the sinking state. So long as they have a queen, or any prospect of raising one, they struggle most vigorously against impending ruin, and never give up, unless their condition is absolutely desperate. In one of my observing hives, I once had a colony of bees, the whole of which might have been spread out on my two hands, busy at work in raising a new queen, from a small piece of brood comb. For two long weeks, they adhered with unfailing perseverance and industry, to their forlorn hope: until at last, one of the two queens which they raised, came forth, and destroyed the other while still in her cell. The bees had now dwindled away to less than half their original number, and the new queen had wings so imperfect that she was unable to fly. I watched their proceedings with great interest; they actually paid very unusual attention to this crippled queen, and treated her more as they are wont to treat a fertile one. In the course of a week, there were not more than a dozen left in the hive, and in a few days more, I missed the queen, and saw only a few disconsolate wretches crawling over the deserted comb! Shame upon the faint-hearted and cowardly of our own race, who, if overtaken by calamity, instead of nobly breasting the dark waters of affliction, and manfully buffetting with their tumultuous waves, meanly resign themselves to their ignoble fate, and sink and perish where they might have lived and triumphed; and double shame upon those who thus "faint in the day of adversity," when living in a Christian land, they might, if they would only receive the word of God, and open the eye of faith, behold a bow of promise spanning the still stormy clouds, and hear a voice bidding them, like the great apostle of the Gentiles, learn not merely to "rejoice in hope of the glory of God," but to "glory in tribulations also."

I have been informed by Mr. Wagner, that Dzierzon has recently devised a plan of *forming nuclei*, substantially the same with my own. His book, however, contemplates having two Apiaries, three or four miles apart, and his plans for multiplying colonies, as there described, were based upon the supposition that the Apiarian will have two such establishments. Such an arrangement would no doubt very greatly facilitate many operations. Our forced swarms might all be removed from the Apiary where they were formed, to the other, and our nuclei treated in the same way, and there would be no necessity for confining the bees after their removal. There are however, weighty objections to such an arrangement, which will prevent it, at least for some time, from being extensively adopted. The labor of removing the bees backwards and forwards, is a serious objection to the whole plan; and in addition to this, the necessity of having a skillful Apiarian at each establishment, puts its adoption out of the question, with most persons who keep bees. It might answer, however, if two bee-keepers, sufficiently far apart, would enter into partnership, and manage their bees as a joint concern. Dzierzon's new plan of creating nuclei, is as follows. Towards evening, remove a piece of brood comb, with eggs and bees just hatching, and put it, with a sufficient number of mature bees, into an empty hive; there must be enough to keep the brood from being chilled over night. If the operation is performed so late that the bees are not disposed to take wing and leave the hive, by morning a sufficient number will have hatched, to supply the place of those which may abandon the nucleus. In my numerous experiments last Summer, in the formation of artificial swarms, I tried this plan and found that it answered a good purpose; the chief objection to it, is the difficulty often of selecting the suitable kind of comb, if the operation is delayed until late in the afternoon. I prefer, therefore, to perform it, when the sun is an hour or two high, and to confine the bees until dark. If there are not a sufficient number of bees on the comb, I shake off some from another frame, directly into the hive, and shut them all up, giving them a supply of water. Sealed queens if possible, should be used in all these operations.

I shall now give a novel mode of creating nuclei, which I have devised, and which I find to be attended with great success. Hive a new swarm in the usual manner, in an old box, and as soon as the bees have entered it, shut them up and carry them down into the cellar. About an hour before sunset, take combs suitable to form as many nuclei as you judge best, say five or six, or even eight or ten if the swarm was large, and you need as many. Bring up the new swarm and shake it out upon a sheet, sprinkling it gently with sugar-water. With a large tumbler or saucer, scoop up without hurting any of the bees, a pint or more of them, and place them before the mouth of one of the hives containing a brood comb; repeat the process, until each nucleus has, say, a quart of bees. If you see the queen, you may give the hive in which you put her, three or four times as many bees as any other; and the next day it may be strengthened with a few combs containing brood, just ready to mature. If you did not find her, at the time of forming the nuclei, when you afterwards examine them, the one which contains her may be properly reinforced with bees and comb, so as to enable it to work to the best advantage.

If this plan of forming nuclei, were attempted earlier in the afternoon it would be difficult to prevent the bees from communicating on the wing, and all going to the hive which contained their queen. If however, the bees when first shaken out of the temporary hive, are so thoroughly sprinkled, as not to be able to take wing and unite together, this mode of forming colonies may be practiced at any hour of the day; and an experienced Apiarian may prefer to do it, as soon as he has fairly hived the new swarm. When the bees are shaken out in front of a hive which has a sealed queen, or eggs from which they can raise one, having a whole night in which to accustom themselves to their new situation, they will be found, the next day, to adhere to the place where they were put, with as much tenacity as a natural swarm does to their new hive. How wonderful that the act of swarming should so thoroughly impress upon the bees, an absolute indisposition to return to the parent stock. If this were a fixed and invariable unwillingness, a sort of blind, unreasoning instinct, it would not be so surprising, but we have already seen that in case the bees lose their queen, they return in a very short time to the stock from which they issued! If the nuclei formed in the manner just described, found in their new hive, no means of obtaining a queen, they would all return, next morning, to the parent stock.

When the Apiarian can obtain a natural swarm from any other Apiary, it may be divided into nuclei in the same way, and even a forced swarm, if brought from a distance, will answer equally well. If the Apiarian wishes to form colonies earlier than the season of natural swarming, and cannot conveniently obtain a forced swarm from an Apiary, at least a mile distant, he may, before the bees begin to fly out in the Spring, transport one of his stocks to a neighbor's, and force from it a swarm at the desired time. Even if it is moved not more than half a mile off, the operation will be almost sure to succeed. Of all modes of forming the nuclei, this I believe will be found to be the neatest, simplest and best.

Having thus described the various ways in which I have successfully formed my nuclei, I shall now show how they may be all built up into powerful stocks. It will be very obvious that on the ordinary plan of management, they would be absolutely worthless, even if it were possible to form them with the common hives. If they were not fed, they would be unable to collect the means of building new comb, and would gradually dwindle away, just as third or fourth swarms which issue late in the season; nor could they be saved even by the most generous feeding, as they would only use their supplies to fill up the little comb they had; so that when the queen was ready to lay, there would be no empty cells to receive her eggs, and too few bees to build any, even if they had all the honey that they required. Such small colonies must gradually waste away, unless they can be speedily and effectually supplied with the requisite number of bees, and this can be done only by hives which give the control of all the combs. With such hives, I can speedily build up my nuclei, (provided I have not formed too many,) to the strength necessary to make them powerful stocks. The hives containing them, ought if possible, to stand at some distance from other hives, say two or three feet: and if this cannot conveniently be done, they should in some way, be so distinguished from the adjoining hives, that the young queens when they are hatched and go out to seek the drones, will not be liable to lose their lives by entering a wrong hive on their return. A small leafy twig fastened on the alighting board of such hives, when they stand near to others, will be almost sure to prevent such a catastrophe: if they stand near to each other, some may be marked in this way, and others with a piece of colored cloth. To guard them against robbers, &c., the entrances to these nuclei should be contracted, so that only a few bees can enter at once. Those which were confined, should be examined, the day after their liberty is given to them; the others, the day after they were formed, when, if they were not supplied with a sealed queen, they will be found actively engaged in constructing royal cells. A new range of comb should now be given to each one, and it should contain no old bees, but brood rapidly maturing, and if possible, eggs and worms only a few days old.

This new addition of strength will greatly encourage the nuclei, and give them the means of starting young queens, if they have not succeeded in doing so with the first comb. I have very frequently found that for some cause which I have not yet ascertained, they often start a large number of queen cells, which in a few days, are all discontinued and untenanted. The second attempt seldom fails. Does practice in this thing make them more expert? But I will simply state the fact, referring to my conjectures on page 218; and remarking that when they make a second attempt, they seem frequently disposed to start a much larger number than they otherwise would have done. In two or three days after giving them the first piece of comb, I give them another, if their queen is nearly mature, and I now let them alone until she ought to be depositing eggs in the hive. I then give them, at intervals of a few days, two or three combs more, and they will now be sufficiently powerful in bees, to gather large quantities of honey, and fill the empty part of their hive. The young queen is supplying with thousands of worker-eggs, the cells from which the brood has emerged, and also the new ones built by the bees, and the new colony will soon be one of the best stock hives in the Apiary. If some of the full frames are moved, and empty ones placed between them, as soon as the bees begin to build powerfully, there need be no guide combs on the empty frames, and still the work will be executed with the most beautiful regularity.

But what, in the mean time, is the condition of the hives from which we are taking so many brood combs for the proper development of our nuclei; are they not weakened so much as to become quite enfeebled? I come now to the very turning point of the whole nucleus system. If due judgment has not been used, and the sanguine bee-keeper has endeavored to multiply his colonies too rapidly, a most grievous disappointment awaits him. Either his nuclei cannot be strengthened at the right time, or this can be done, only by impoverishing the old stocks, and the result of the whole operation will be a most decided failure, and if he is in the vicinity of sugar-houses, confectionaries, or other tempting places of bee resort, he will find the population of his colonies very seriously diminished, and will have to break up the most of the nuclei which he had formed, and incur the danger of losing nearly the whole of his stock. I lay it down as a fundamental principle in my nucleus system, that the old stocks must never be so much weakened by the removal of brood-comb and bees that they are not able to keep their numbers sufficiently strong to refill rapidly all the vacancies among their combs. If the Apiarian attempts to multiply his stocks so rapidly that this cannot be done, I will ensure him ample cause to repent at leisure of his folly. If however, the attempt at very rapid multiplication is made only by those who are favorably situated, and who have skill in the management of bees, a very large gain may be made in the number of stocks, and they may all be strong and flourishing.

If a strong stock of bees in a hive of moderate size, which admits of thorough inspection, is examined at the height of the honey harvest, nearly all the cells will often be found filled with brood, honey or bee-bread. The great laying of the queen, according to some writers, is now over, not however as they erroneously imagine, because her fertility has decreased, but merely because there is not *room* in the hive for all her eggs. She may often be seen restlessly traversing the combs, seeking in vain for empty cells, until finding none, she is compelled to extrude her eggs only to be devoured by the bees. If some of the full combs are removed, and empty ones substituted in their place, she will speedily fill them, laying at the rate of two or three thousand a day! When my strong stocks are from time to time deprived of one or two combs, if honey can easily be procured,[19] the bees proceed at once to replace them, and the queen commences laying in the new combs as soon as the cells are fairly started. If the combs are not removed *too fast*, and care is taken not to deprive the stock of so much brood that the bees cannot keep up a vigorous population, a queen in a hive so managed, will lay her eggs in cells to be nurtured by the bees, instead of being eaten up; and thus, in the course of the season, she may become the mother of three or four times as many bees, as are reared in a hive under other circumstances. By careful management, brood enough may, in this way, be taken from a single hive, to build up a large number of nuclei. Towards the close of the season however, as such a hive has been constantly tasked in building comb and feeding young bees, almost all its honey will have been used for these purposes, and although it may be very populous, unless it is liberally fed, it will be sure to perish. Since the discovery that unbolted rye flour will answer so admirably as a substitute for pollen, we can supply the bees not only with honey, when none can be obtained from the blossoms, but with an abundance of bee-bread, when pollen is scarce. As I am writing this chapter, (March 29, 1853,) my bees are zealously engaged in taking the flour from some old combs in front of their hives, and they can be seen most beautifully moulding the little pellets on their thighs. By my movable combs, I can give them the flour at once in their hives, as it can easily be rubbed into an empty comb. The importance of Dzierzon's discovers of a substitute for pollen, can hardly be over-estimated. If he had done nothing more for the cause of Apiarian science, no true-hearted bee-keeper would ever allow his name to be forgotten.

In the Chapter on Feeding, I shall give more specific directions as to the way in which the cultivator must feed his bees, when he aims at increasing, as rapidly as possible, the number of his stocks. Unless this work is done with great judgment, he will find often that the more he feeds, the less bees he has in his hives, the cells being all occupied with honey instead of brood. Such is the passion of bees for storing away honey, that large supplies of it will always most seriously interfere with breeding, unless the bees are sufficiently numerous to build new comb in which the queen can find room for her eggs.

I have no doubt that some who have but little experience in the management of bees, are ready to imagine that they could easily strike out a simpler and better way of increasing the number of colonies. For instance: let a full hive have half its comb and bees put into an empty hive, and the work of doubling, is without further trouble, effectually accomplished. But what will the queenless hive do, under such circumstances? Why, build of course, queen cells, and rear another. But what kind of comb will they fill their hives with, before the young queen begins to breed?

Of that, perhaps, you had never thought. Let me now lay down the only safe rule for all who engage in the multiplication of artificial swarms. Never, under *any* circumstances, take so much comb and brood from your stock hives, as seriously to reduce their numbers. This should be to the Apiarian, as "the law of the Medes and Persians, which altereth not."

Suppose that I divide a populous stock, about swarming season, into four or five colonies; the strong probability is, that not one of them, if left to themselves, will be strong enough to survive the Winter. If fed in the ordinary way, and yet not supplied with combs and bees, their ruin will often be only accelerated. If, on the contrary, I had taken, from time to time, combs sufficient to form three or four nuclei, and had strengthened the new colonies, in such a way as not to draw too severely upon the resources of the parent stock, I might expect to see them all, in due time, strong and flourishing.

In the Spring of the year, if I desire to determine the strength of a colony principally to raising young bees, I can easily effect it by the following plan. A box is made, of the same inside dimensions with the lower hive, into which the combs and bees of a full hive can all be transferred, as soon as the bees are gathering honey enough to build new combs. This box is now set over the old hive, which contains its complement of frames with guide combs, or better still, with empty combs. As soon as the bees begin to build, they take possession of the lower hive, through which they go in and out, and the queen descends with them, in order to lay her eggs in the lower combs. As soon as the old apartment becomes pretty well filled, a large number of combs with maturing bees, may be taken from the upper one, and when the hive below is full, they may all be safely removed. If none of the upper combs are removed, they will be filled with honey, as soon as the brood is hatched; and as they will contain large stores of bee-bread, they will answer admirably for replenishing stocks which have an insufficient supply. In no other way, so far as I know, can so much honey be secured, and if quantity, not quality, is aimed at, or if the test of quality is its fitness for the use of the bees, I would recommend this mode as superior to any other. If two swarms are hived together, or a very powerful stock is lodged in a hive, so that at once they can have access to the upper apartment, an extraordinary quantity of honey can be secured, and of a very excellent quality. As soon as the bees have raised one generation of young, in the combs of the upper box, or rather in a part of them, they will use it chiefly for storing honey, and all that it contains may be taken from them. In flavor, it will be found to be nearly as good as honey stored in what is called "virgin comb."

In the Chapter on the Requisites of a good hive, I have said that in size it should be adapted to the natural instincts of the bee, and yet admit of enlarging or contracting, according to the wants of the colony placed in it. I never use a hive, the main apartment of which, holds less than a Winchester bushel. If small colonies are placed in such a hive, it must be temporarily partitioned off, to suit the size of its inmates; for if bees have too much room given to them, they cannot concentrate their animal heat, and are so much discouraged that they often abandon their hive. I am aware that many judicious Apiarians recommend hives of much smaller dimensions, and I shall now give my reasons for using one so large. If a hive is too small, then in the Spring, the combs are soon filled with honey, bee-bread and brood, and the surprising fertility of the queen bee, can be turned to no efficient account. If the honey-harvest in any year, is deficient, such a colony is very apt to perish in the succeeding Winter; whereas in a large hive, the honey stored up in a fruitful season, is a reserve supply, in time of need. In very large hives, I have seen large accumulations of honey which have been untouched for years, while on the same stand, stocks of about the same age, in small hives have perished by starvation. A good early swarm in any situation at all favorable, will fill, the first season, a hive that holds a bushel: and if there is any location in which they cannot do this, a doubled swarm should be put into the hive, or bee-keeping may, as far as profit is concerned, be abandoned. But it may be objected that if the swarm was not sufficiently strong to fill their hive, the bees often suffer from the cold in Winter, and become too much reduced in numbers, to build early and rapidly in the ensuing Spring. This is undoubtedly true, and hence the great importance of putting a generous allowance of bees into a hive at the first start, unless, as on my plan, the requisite strength can be given to them, at a subsequent period. The hive, if large, should be all the more carefully protected from extremes of cold, in order to give the bees an opportunity of developing their natural powers of re-production, to the best advantage.

In such a hive, the queen will be able to breed almost every month in the year, even in the coldest climates where bees flourish, and on the return of Spring, thousands of young bees will be found in it, which could not have been bred in a small, or badly protected hive. The Polish hives described by Mr. Dohiogost, have already been referred to. Some of these hold about three bushels, and yet the bees swarm from them with great regularity, and the swarms are often of immense size. These hives are admirably protected, and at the time of hiving at least *four* times the number of bees are lodged in them, that are ordinarily put into one of our hives. The queen bee, in such a hive, has ample room to lay her three thousand eggs, or more, daily: and a prodigious colony is raised, which often stores enormous supplies of honey. As all the frames in my hives are of the same dimensions, the size of the hive may be conveniently varied, to suit the views of different bee-keepers; for they may be large or small, according to the number of frames designed to be used. I hope, before long, to experiment with hives as large again, as those that I now use; or rather, with such, as by containing an upper box, may be made to accommodate twice as many bees. This whole subject of the proper size of hives, certainly needs to be taken entirely out of the region of conjecture, and to be put upon the basis of careful observations. Unquestionably the size will require, in some respects, to be modified by the more or less favorable character of the country for bee-keeping; but I am satisfied that small hives will be found of but little profit, and that large ones, unless well stocked with bees, from the first, and thoroughly protected, will often fail to answer any good end. If I should find on further experiment, that the very large hives of which I have spoken, are better, my hives are at present so constructed that without any alteration of existing parts, they can easily be supplied with the required additions. I have already mentioned that I sometimes build my hives, three in one structure, in order to save expense in their construction. I do not however, wish to be considered as recommending such hives as the best for general use. For some purposes a single hive is unquestionably the best, as it can be easily moved by one person; and this, will many times be found to be a point of great importance. The double hives, or two in one, are for most purposes, decidedly the best, as well as the cheapest. I have quite recently contrived a plan of constructing my wooden hives in such a manner as to give them very great protection against extremes of heat and cold, while at the same time they can be easily and cheaply made, by any one who can handle the simplest mechanical tools.

It has been previously stated that the queen bee cannot be induced to sting, by any kind of treatment however severe. The reason of this strange unwillingness to use her natural and powerful weapon, will be obvious, when we consider how indispensable the preservation of her life is to the very existence of the colony, and that her single sting, the loss of which would be her death, could avail but little for their defence, in case of an attack. She never uses her weapon, except when engaged in mortal combat with another queen. As soon as the two rivals come together, they clinch, at once, with every demonstration of the most vindictive hatred. Why then, are not both of them often destroyed? and why are not hives, in the swarming season, almost certain to become queenless? We can never sufficiently admire the provision so simple and yet so effectual, by which such a calamity is prevented. The queen bee never stings unless she has such an advantage in the combat, that she can curve her body under that of her rival, in such a manner as to inflict a deadly wound, without any risk of being stung herself! The moment that the position of the two combatants is such that neither has the advantage, and that both are liable to perish, they not only refuse to sting, but disengage themselves, and suspend their conflict for a short time! If it were not for this peculiarity of instinct, such combats would very often terminate in the death of both the parties, and the race of bees would be in danger of becoming extinct.

The unwillingness of a swarm of bees, which has been deprived of its queen, to receive another, until after some time has elapsed, must always be borne in mind, by those who have anything to do with making artificial swarms. About 24 hours must elapse before it will be safe to introduce a strange mother into a queenless hive; and even then, if she is not fertile, she will run a great risk of being destroyed. To prevent such losses, I adopt the German plan of confining the queen, in what they call, "a queen cage." A small hole, about as large as a thimble, may be gouged out of a block, and covered over with wire gauze, or any other kind of perforated cover, so that when the queen is put in, the bees cannot enter to destroy her. Before long, they will cultivate an acquaintance, by thrusting their antennæ through to her; so that, when she is liberated the next day, they will gladly adopt her in place of the one they have lost. If a hole large enough for her to creep out, is closed with wax, they will gnaw the wax away, and liberate her themselves, from her confinement. Queens that seem bent on departing to the woods, may be confined in the same way, until the colony has given up all thoughts of forsaking its hive. A small paste-board box with suitable holes, or a wooden match-box thoroughly scalded, I have found to answer a very good purpose.

I shall here describe what may be called a *Queen Nursery* which I have contrived to aid those who are engaged in the rapid multiplication of colonies by artificial means. A solid block about an inch and a quarter thick, is substituted for one of my frames; holes, about one and a half inches in diameter, are bored through it, and covered on both sides, with gauze wire slides; the wire ought to be such as will allow a common bee to pass through, but should be too small to permit a queen to do the same. Any kind of perforated cover may be made to answer the same purpose as the gauze wire. If a number of sealed queens are on hand, and there is danger that some may hatch, and destroy the others, before the Apiarian can make use of them in forming artificial swarms, he may very carefully cut out the combs containing them, and place them each in a separate cradle! The bees having access to them, will give them proper attention, and as soon as they are hatched, will supply them with food, and thus they will always be on hand for use when they are needed. This Nursery must of course, be established in a hive which has no mature queen, or it will quickly be transformed into a slaughter house by the bees. I have not yet tested this plan so thoroughly as to be *certain* that it will succeed; and I know so well the immense difference between theoretical conjectures and practical results, that I consider nothing in the bee line, or indeed in any other line, as established, until it has been submitted to the most rigorous demonstrations, and has triumphantly passed from the mere regions of the brain to those of actual fact. A theory on any subject may seem so plausible as almost to amount to a positive demonstration, and yet when put to the working test, it is often found to be encumbered by some unforeseen difficulty, which speedily convinces even its sanguine projector, that it has no practical value. Nine things out of ten may work to a charm, and yet the tenth may be so connected with the other nine, that its failure renders their success of no account. When I first used this Nursery, I did not give the bees access to it, and I found that the queens were not properly developed, and died in their cells. Perhaps they did not receive sufficient warmth, or were not treated in some other important respects, as they would have been if left under the care of the bees. In the multiplicity of my experiments, I did not repeat this one under a sufficient variety of circumstances, to ascertain the precise cause of failure; nor have I as yet, tried whether it will answer perfectly, by admitting the bees to the queen cells.

Last Spring, I made one queen supply several hives with eggs, so as to keep them strong in numbers while they were constantly engaged in rearing a large number of spare queens. Two hives which I shall call

A and B, were deprived at intervals of a week, each of its queen,[20] in order to induce them to raise a number of young sealed queens for the use of the Apiary. As soon as the queens in A, were of an age suitable to be removed, I took them away and gave the colony a fertile queen from another hive, C; as soon as she had laid a large number of eggs in the empty cells, I removed the queen cells now sealed over, from B, and gave them the loan of this fertile mother, until she had performed the same necessary office for them. By this time, the queen cells in C, were sealed over; these were now removed, and the queen restored; she had thus made one circuit, and laid a very large number of eggs in the two hives which were first deprived of their queens. After allowing her to replenish her own hive with eggs, I sent her out again on her perambulating mission, and by this new device was able to get an extraordinary number of young queens from the three hives, and at the same time to preserve their numbers from seriously diminishing. Two queens may in this way, be made in six hives to furnish all the supernumerary queens which will be wanted in quite a large Apiary.

It will be perfectly obvious to every intelligent and ingenious Apiarian, that the perfect control of the comb, is the *soul* of an entirely new system of practical management, and that it may be modified to suit the wants of all who wish to cultivate bees. Even the advocate of the old fashioned plan of killing the bees, can with one of my hives, destroy his faithful laborers, by shaking them into a tub of water, almost, if not quite as speedily as by setting them over a sulphur pit; while after the work of death is accomplished, his honey will be free from disgusting fumes, and all the labor of cutting it out of the hive, may be dispensed with.

I am now prepared to answer an objection which doubtless has been present in the minds of many, all the time that they have been reading the various processes on which I rely for the multiplication of colonies. A very large number of persons who keep bees, or who wish to keep them, are so much afraid of them that they object entirely even to natural swarming, because they are in danger of being stung in the process of hiving the bees. How are such persons to manage bees on my plan, which seems like bearding a lion in its very den! The truth is that some persons are so very timid, or suffer so dreadfully from the sting of a bee, that they are every way disqualified from having anything to do with them, and ought either to have no bees upon their premises, or to entrust the care of them to some suitable person. By managing bees according to the directions furnished in this treatise, almost any one can learn, by using a bee-dress, to superintend them, with very little risk; while those who are favorites with them, may dispense entirely with any protection. I find in short, that the risk of being stung is really diminished by the use of my hives; although it will be hard to convince those who have not seen them in use, that this can be so.

There is still another class who either keep bees or can be induced to keep them, and who are anxiously inquiring for some new hive or new plan by which, with little or no trouble, they may reap copious harvests of the precious nectar. This is emphatically *the* class to seize hold of every new device, and waste their time and money to fill the coffers of the ignorant or unprincipled. There never will be a "royal road" to profitable bee-keeping. If there is any branch of rural economy which more than all others demands care and experience, for its profitable management, it is the keeping of bees; and those who have a painful consciousness that the disposition to put off and neglect, was, so to speak, born in them, and has never been got out of them, will do well to let bees alone, unless they hope, by the study of their systematic industry, to reform evil habits which are well nigh incurable.

While I feel very sanguine that my system of management will be used extensively and very advantageously, by careful and skillful Apiarians, I know too much of the world to expect that it will, with the masses, very speedily supercede other methods, even if it were so absolutely perfect, as to admit of no possible improvement. I hope, however, that I may, without being charged with presumption, be permitted to put on record the prediction, that *movable frames* will in due season, be almost universally employed; and this, whether bees are allowed to swarm naturally, or are increased by artificial means, or are kept in hives in which they are not expected to swarm at all.

Note.—The very day on which I first contrived the plan, so perfectly simple, and yet so efficacious, of gaining the control of the combs
by these frames, I not only foresaw all the consequences which would
follow their adoption, but wrote as follows, in my Bee-Journal. "The use
of these frames will, I am persuaded, give a new impulse to the easy and
profitable management of bees; and will render the making of artificial
swarms an easy operation."

CHAPTER XII. THE BEE-MOTH, AND OTHER ENEMIES OF BEES. DISEASES OF BEES.

O f all the numerous enemies of the honey-bee, the Bee-Moth (Tinea mellonella,) in climates of hot Summers, is by far, the most to be dreaded. So wide spread and fatal have been its ravages in this country, that thousands have abandoned the cultivation of bees in despair, and in districts which once produced abundant supplies of the purest honey, bee-keeping has gradually dwindled down into a very insignificant pursuit. Contrivances almost without number, have been devised, to defend the bees against this invidious foe, but still it continues its desolating inroads, almost unchecked, laughing as it were to scorn, at all the so-called "moth-proof" hives, and turning many of the ingenious fixtures designed to entrap or exclude it, into actual aids and comforts in its nefarious designs.

I should feel but little confidence in being able to reinstate bee-keeping in our country, into a certain and profitable pursuit, if I could not show the Apiarian in what way he can safely bid defiance to the pestiferous assaults of this, his most implacable enemy. I have patiently studied its habits for years, and I am at length able to announce a system of management founded upon the peculiar construction of my hives, which will enable the careful bee-keeper to protect his colonies against the monster. The careful bee-keeper, I say: for to pretend that the careless one, can by any contrivance effect this, is "a snare and a delusion;" and no well-informed man, unless he is steeped to the very lips, in fraud and imposture, will ever claim to accomplish any thing of the kind. The bee-moth infects our Apiaries, just as weeds take possession of a fertile soil; and the negligent bee-keeper will find a "moth-proof" hive, when the sluggard finds a *weed-proof* soil, and I suspect not until a consummation so devoutly wished for by the slothful has arrived. Before explaining the means upon which I rely, to circumvent the moth, I will first give a brief description of its habits.

Swammerdam, towards the close of the 17th century, gave a very accurate description of this insect, which was then called by the very expressive name of the "bee-wolf." He has furnished good drawings of

it, in all its changes, from the worm to the perfect moth, together with
the peculiar webs or galleries which it constructs and from which the
name of Tinea Galleria or gallery moth, has been given to it by some
entomologists. He failed, however, to discriminate between the male
and female, which, because they differ so much in size and appearance,
he supposed to be two different species of the wax-moth. It seems to
have been a great pest in his time; and even Virgil speaks of the "dirum
tineæ genus," the dreadful *offspring* of the moth; that is the worm. This
destroyer usually makes its appearance about the hives, in April or May;
the time of its coming, depending upon the warmth of the climate, or
the forwardness of the season. It is seldom seen on the wing, (unless
startled from its lurking place about the hive,) until towards dark, and is
evidently, chiefly nocturnal in its habits. In dark cloudy days, however, I
have noticed it on the wing long before sunset, and if several such days
follow in succession, the female oppressed with the urgent necessity of
laying her eggs, may be seen endeavoring to gain admission to the hives.
The female is much larger than the male, and "her color is deeper and
more inclining to a darkish gray, with small spots or blackish streaks on
the interior edge of her upper wings." The color of the male inclines more
to a light gray; they might easily be mistaken for different species of moths.
These insects are surprisingly agile, both on foot and on the wing. The
motions of a bee are very slow in comparison. "They are," says Reaumur,
"the most nimble-footed creatures that I know." "If the approach to the
Apiary[21] be observed of a moonlight evening, the moths will be found
flying or running round the hives, watching an opportunity to enter,
whilst the bees that have to guard the entrances against their intrusion,
will be seen acting as vigilant sentinels, performing continual rounds near
this important post, extending their antennæ to the utmost, and moving
them to the right and left alternately. Woe to the unfortunate moth that
comes within their reach!" "It is curious," says Huber, "to observe how
artfully the moth knows how to profit, to the disadvantage of the bees,
which require much light for seeing objects; and the precautions taken
by the latter in reconnoitering and expelling so dangerous an enemy."

The entrance of the moth into a hive, and the ravages committed by her progeny, forcibly remind one of the sad havoc which sin often
makes of character and happiness, when it finds admission into the
human heart, and is allowed to prey unchecked, upon all its most precious
treasures; and he who would not be so enslaved by its power, as to lose

all his spiritual life and prosperity, must be constantly on the defensive, and ever on the "watch" against its fatal intrusions.

Only some tiny eggs are deposited by the moth, and they give birth to a very delicate, innocent-looking worm; but let these apparently insignificant creatures once "get the upper hand," and all the fragrance of the honied dome, is soon corrupted by their abominable stench; every thing beautiful and useful, is ruthlessly destroyed; the hum of happy industry is stilled, and at last, nothing is left in the desecrated hive, but a set of ravenous, half famished worms, knotting and writhing around each other, in most loathsome convolutions.

Wax is the proper aliment of the larvæ of the bee-moth: and upon this seemingly indigestible substance, they thrive and fatten. When obliged to steal their living as best they can, among a powerful stock of bees, they are exposed, during their growth, to many perils, and seldom fare well enough to reach their natural size: but if they are rioting at pleasure, among the full combs of a feeble and discouraged population, they often attain a size and corpulency truly astonishing. If the bee-keeper wishes to see their innate capabilities fully developed, let him rear a lot for himself among some old combs, and if prizes were offered for fat and full grown worms, he might easily obtain one. In the course of a few weeks, the larva like that of the silk-worm, stops eating, and begins to think of a suitable place for encasing itself in its silky shroud. In hives where they reign uncontrolled, this is a work of but little difficulty; almost any place will answer their purpose, and they often pile their cocoons, one on top of another, or join them in long rows together: but in hives strongly guarded by healthy bees, this is a matter not very easily accomplished; and many a worm while it is cautiously prying about, to see where it can find some snug place in which to ensconce itself, is caught by the nape of the neck, and very unceremoniously served with an instant writ of ejection from the hive. If a hive is thoroughly made, of sound materials, and has no cracks or crevices under which the worm can retreat, it is obliged to leave the interior in search of such a place, and it runs a most dangerous gantlet, as it passes, for this purpose, through the ranks of its enraged foes. Even in the worm state, however, its motions are exceedingly quick; it can crawl backwards or forwards, and as well one way as another: it can twist round on itself, curl up almost into a knot, and flatten itself out like a pancake! in short, it is full of stratagems and cunning devices. If obliged to leave the hive, it gets under any board or concealed crack, spins its cocoon, and patiently awaits its transformation. In most of the common hives, it is under no necessity of leaving its birth place for this purpose. It is almost certain to find a crack or flaw into which it can creep, or a small space between the bottom-board and the edges of the hive which rest upon it. A *very* small crevice will answer all its purposes. It enters, by flattening itself out almost as much as though it had been passed under a roller, and as soon as it is safe from the bees, it speedily begins to give its cramped tenement, the requisite proportions. It is utterly amazing how an insect apparently so feeble, can do this; but it will often gnaw for itself a cavity, even in solid wood, and thus enlarge its retreat, until it has ample room for making its cocoon! The time when it will break forth into a winged insect, depends entirely upon the degree of heat to which it is exposed. I have had them spin their cocoons and hatch in a temperature of about 70°, in ten or eleven days, and I have known them to spin so late in the Fall, that they remained all Winter, undeveloped, and did not emerge until the warm weather of the ensuing Spring!

If they are hatched in the hive, they leave it, in order to attend to the business of impregnation. In the moth state, they do not actually attack the hives, to plunder them of food, although they have a "sweet tooth" in their head, and are easily attracted by the odor of liquid sweets. The male, having no special business in the hive, usually keeps himself at a safe distance from the bees: but the female, impelled by an irresistible instinct, seeks admission, in order to deposit her eggs where her offspring may gain the readiest access to their natural food. She carefully explores all the cracks and crevices about the bottom-board, and if she finds a suitable place under them, lays her eggs among the parings of the combs, and other refuse matter which has fallen from the hive. If she enters a feeble or discouraged stock, where she can act her own pleasure, she will lay her eggs among the combs. In a hive where she is too closely watched to effect this, she will insert them in the corners, into the soft propolis, or in any place where there are small pieces of wax and bee-bread, which have fallen upon the bottom-board, and which will furnish a temporary place of concealment for her progeny, and also the requisite nourishment, until they have strength and enterprise enough to reach the main combs of the hive, and fortify themselves there. "As soon as hatched,[22] the worm encloses itself in a case of white silk, which it spins around its body; at first it is like a mere thread, but gradually increases in size, and during its growth, feeds upon the cells around it, for which purpose it has only to put forth its head, and find its wants supplied. It devours its food with great avidity, and consequently increases so much in bulk, that its gallery soon becomes too short and narrow, and the creature is obliged to thrust itself forward and lengthen the gallery, as well to obtain more room as to procure an additional supply of food. Its augmented size exposing it to attacks from surrounding foes, the wary insect fortifies its new abode with additional strength and thickness, by blending with the filaments of its silken covering, a mixture of wax and its own excrement, for the external barrier of a new gallery, the *interior* and partitions of which are lined with a smooth surface of white silk, which admits the occasional movements of the insect, without injury to its delicate (?) texture. In performing these operations, the insect might be expected to meet with opposition from the bees, and to be gradually rendered more assailable as it advanced in age. It never, however, exposes any part but its head and neck, both of which are covered with stout helmets or scales impenetrable to the sting of a bee, as is the composition of the galleries that surround it." As soon as it has reached its full growth, it seeks in the manner previously described, a secure place for undergoing its changes into a winged insect.

Before describing the way in which I protect my hives from this deadly pest, I shall first show why the bee-moth has so wonderfully increased in numbers in this country, and how the use of patent hives has so powerfully contributed to encourage its ravages. It ought to be borne in mind that our climate is altogether more propitious to its rapid increase, than that of Great Britain. Our intensely hot summers develop most rapidly and powerfully, insect life, and those parts of our country where the heat is most protracted and intense, have, as a general thing, suffered most from the devastations of the bee-moth.

The bee is not a native of the American continent; it was first brought here by colonists from Great Britain, and was called by the Indians, the white man's fly. With the bee, was introduced its natural enemy, created for the special purpose, not of destroying the insect, on whose industry it thrives, and whose extermination would be fatal to the moth itself, but that it might gain its livelihood as best it could in this busy world. Finding itself in a country whose climate is exceedingly propitious to its rapid increase, it has multiplied and increased a thousand fold, until now there is hardly a spot where the bees inhabit, which is not infested by its powerful enemy.

I have often listened to the glowing accounts of the vast supplies of honey obtained by the first settlers, from their bees. Fifty years ago, the markets in our large cities were much more abundantly supplied than they now are, and it was no uncommon thing to see exposed for sale, large washing-tubs filled with the most beautiful honey. Various reasons have been assigned for the present depressed state of Apiarian pursuits. Some imagine that newly settled countries are most favorable for the labors of the bee: others, that we have overstocked our farms, so that the bees cannot find a sufficient supply of food. That neither of these reasons will account for the change, I shall prove more at length, in my remarks on Honey, and when I discuss the question of overstocking a district with bees. Others lay all the blame upon the bee-moth, and others still, upon our departure from the good old-fashioned way of managing bees. That the bee-moth has multiplied most astonishingly, is undoubtedly true. In many districts, it so superabounds, that the man who should expect to manage his bees with as little care as his father and grandfather bestowed upon them, and yet realize as large profits, would find himself most wofully mistaken. The old bee-keeper often never looked at his bees after the swarming season, until the time came for appropriating their spoils. He then carefully "hefted" all his hives so as to be able to judge as well as he could, how much honey they contained. All which were found to be too light to survive the Winter, he at once condemned; and if any were deficient in bees, or for any other reason, appeared to be of doubt-ful promise, they were, in like manner, sentenced to the sulphur pit. A certain number of those containing the largest supplies of honey, were also treated in the same summary way: while the requisite number of the *very best*, were reserved to replenish his stock another season. If the same system precisely, were now followed, a number of colonies would still perish annually, through the increased devastations of the moth.

The change which has taken place in the circumstances of the bee-keeper, may be illustrated by supposing that when the country was first settled, weeds were almost unknown. The farmer plants his corn, and then lets it alone, and as there are no weeds to molest it, at the end of the season he harvests a fair crop. Suppose, however, that in process of time, the weeds begin to spread more and more, until at last, this farmer's son or grandson finds that they entirely choke his corn, and that he cannot, in the old way, obtain a remunerating crop. Now listen to him, as he gravely informs you that he cannot tell how it is, but corn with him has all "run out." He manages it precisely as his father or grandfather always managed theirs, but somehow the pestiferous weeds will spring up, and he has next to no crop. Perhaps you can hardly conceive of such transparent ignorance and stupidity; but it would be difficult to show that it would be one whit greater than that of a large number who keep bees in places where the bee-moth abounds, and who yet imagine that those plans which answered perfectly well fifty or a hundred years ago, when moths were scarce, will answer just as well now.

If however, the old plan had been rigidly adhered to, the ravages of the bee-moth would never have been so great as they now are. The introduction of *patent hives* has contributed most powerfully, to fill the land with the devouring pest. I am perfectly aware that this is a bold assertion, and that it may, at first sight, appear to be very uncourteous, if not unjust, to the many intelligent and ingenious Apiarians, who have devoted much time, and spent large sums of money, in perfecting hives designed to enable the bee-keeper to contend most successfully against his worst enemy. As I do not wish to treat such persons with even the appearance of disrespect, I shall endeavor to show just how the use of the hives which they have devised, has contributed to undermine the prosperity of the bees. Many of these hives have valuable properties, and if they were always used in strict accordance with the enlightened directions of those who have invented them, they would undoubtedly be real and substantial improvements over the old box or straw hive, and would greatly aid the bee-keeper in his contest with the moth. The great difficulty is that they are none of them, able to give him the facilities which alone can make him victorious. No hive, as I shall soon show, can ever do this, which does not give the complete and easy control of all the combs.

I do not know of a single improved hive which does not aim at entirely doing away with the old-fashioned plan of killing the bees. Such a practice is denounced as being almost as cruel and silly as to kill a hen for the sake of obtaining her feathers or a few of her eggs. Now if the Apiarian can be furnished with suitable instructions, and such as he will *practice*, for managing his bees so as to avoid this necessity, then I admit the full force of all the objections which have been urged against it. I have never read the beautiful verses of the poet Thompson, without feeling all their force:

"Ah, see, where robbed and murdered in that pit Lies the still heaving hive! at evening snatched, Beneath the cloud of guilt-concealing night, And fixed o'er sulphur! while, not dreaming ill, The happy people, in their waxen cells, Sat tending public cares; Sudden, the dark oppressive steam ascends, And, used to milder scents, the tender race, By thousands, tumble from their honied dome! Into a gulf of blue sulphureous flame."

The plain matter of fact however, is, that in our country, as many bees, if not more, die of starvation in their hives, as ever were killed by the fumes of sulphur. Commend me rather to the humanity of the old-fashioned bee-keeper, who put to a speedy and therefore merciful death, the poor bees which are now, by millions, tortured by slow starvation among their empty combs! At the present time, (April 1853,) I am almost daily hearing of swarms which have perished in this way, during the last Winter; and I know of only one person who was merciful enough to kill his weak stocks, rather than suffer them to die so cruel a death.

If the use of the common patent hives could only keep the stocks strong in numbers, and if the bee-keepers would always see that they were well supplied with honey, then I admit that to kill the bees would be both cruel and unnecessary. Such however, are the discouragements and losses necessarily attending the use of any hive which does not give the control of the combs, that there will be few who do not continually find that some of their stocks are too feeble to be worth the labor and expense of attempting to preserve them over Winter. How many colonies are annually wintered, which are not only of no value to their owner, but are positive nuisances in his Apiary; being so feeble in the Spring, that they are speedily overcome by the moth, and answer only to breed a horde of destroyers to ravage the rest of his Apiary. The time spent upon them is often as absolutely wasted, as the time devoted to a sick animal incurably diseased, and which can never be of any service, while by nursing it along, its owner incurs the risk of infecting his whole stock with its deadly taint. If, on the score of kindness, he should shut it up, and let it starve to death, few of us, I imagine, would care to cultivate a very intimate acquaintance with one so extremely original in the exhibition of his humanity!

Ever since the introduction of patent hives, the notion has almost universally prevailed, that stocks must not, under *any* circumstances, be voluntarily broken up; and hence, instead of Apiaries, filled in the Spring, with strong and healthy stocks of bees, easily able to protect themselves against the bee-moth, and all other enemies, we have multitudes of colonies which, if they had been kept on purpose to furnish food for the worms, could scarcely have answered a more valuable end in encouraging their increase. The simple truth is, that improved hives, without an improved system of management, have done on the whole more harm than good; in no country have they been so extensively used as in our own, and no where has the moth so completely gained the ascendency. Just so far as they have discouraged bee-keepers from the old plan of killing off all their weak swarms in the Fall, just so far have they extended "aid and comfort" to the moth, and made the condition of the bee-keeper worse than it was before. That some of them might be managed so as in all ordinary cases, to give the bees complete protection against their scourge, I do not, for a moment, question; but that they cannot, from the very nature of the case, answer fully in all emergencies, the ends for which they were designed, I shall endeavor to prove and not to assert.

The kind of hives of which I have been speaking, are such as have been devised by intelligent and honest men, practically acquainted with the management of bees: as for many of the hives which have been introduced, they not only afford the Apiarian no assistance against the inroads of the bee-moth, but they are so constructed as positively to aid it in its nefarious designs. The more they are used, the worse the poor bees are off: just as the more a man uses the lying nostrums of the brazen-faced quack, the further he finds himself from health and vigor.

I once met with an intelligent man who told me that he had paid a considerable sum, to a person who professed to be in possession of many valuable *secrets* in the management of bees, and who promised, among other things, to impart to him an infallible remedy against the bee-moth. On the receipt of the money, he very gravely told him that the secret of keeping the moth out of the hive, was to keep the bees strong and vigorous! A truer declaration he could not have made, but I believe that the bee-keeper felt, notwithstanding, that he had been imposed upon, as outrageously, as a poor man would be, who after paying a quack a large sum of money for an infallible, life-preserving secret, should be turned off with the truism that the secret of living forever, was to keep well!

There is not an intelligent, observing Apiarian who has been in the habit of carefully examining the operations of bees, not only in his own Apiary, but wherever he could find them, who has not seen strong stocks flourishing under almost any conceivable circumstances. They may be seen in hives of the most miserable construction, unpainted and unprotected, sometimes with large open cracks and clefts extending down their sides, and yet laughing to defiance, the bee-moth, and all other adverse influences.

Almost any thing hollow, in which the bees can establish them-selves, and where they have once succeeded in becoming strong, will often be successfully tenanted by them for a series of years. To see such hives, as they sometimes may be seen, in possession of persons both ignorant and careless, and who hardly know a bee-moth from any other kind of moth, may at first sight well shake the confidence of the inquirer, in the necessity or value of any particular precautions to preserve his hives from the devastations of the moth.

After looking at these powerful stocks in what may be called log-cabin hives, let us examine some in the most costly hives, which have ever been constructed; in what have been called real "Bee-Palaces;" and we shall often find them weak and impoverished, infested and almost devoured by the worms. Their owner, with books in his hand, and all the newest devices and appliances in the Apiarian line, unable to protect his bees against their enemies, or to account for the reason why some hives seem, like the children of the poor, almost to thrive upon ill-treatment and neglect, while others, like the offspring of the rich and powerful, are feeble and diseased, almost in exact proportion to the means used to guard them against noxious influences, and to minister most lavishly to all their wants.

I once used to be much surprised to hear so many bee-keepers speak of having "good luck," or "bad luck" with their bees; but really as bees are generally managed, success or failure does seem to depend almost entirely upon what the ignorant or superstitious are wont to call "luck."

I shall now try to do what I have never yet seen satisfactorily done by any writer on bees; viz.: show exactly under what circumstances the bee-moth succeeds in establishing itself in a hive; thus explaining why some stocks flourish in spite of all neglect, while others, in the common hives, fall a prey to the moth, let their owner be as careful as he will, I shall finally show how in suitable hives, with proper precautions, it may always be kept from seriously annoying the bees.

It often happens, when a large number of stocks are kept, that in spite of all precautions, some of them are found in the Spring, so greatly reduced in numbers, that if left to themselves, they are in danger of fall-ing a prey to the devouring moth. Bees, when in feeble colonies, seem often to lose a portion of their wonted vigilance, and as they have a large quantity of empty comb which they cannot guard, even if they would, the moth enters the hive, and deposits a large number of eggs, and thus before the bees have become sufficiently numerous to protect themselves, the combs are filled with worms, and the destruction of the colony speedily follows. The ignorant or careless bee-keeper is informed of the ravages which are going on in such a hive, only when its ruin is fully completed, and a cloud of winged pests issues from it, to destroy if they can, the rest of his stocks. But how, it may be asked, can it be ascertained that a hive is seriously infested with the all-devouring worms? The aspect of the bees, so discouraged and forlorn, proclaims at once that there is trouble of some kind within. If the hive be slightly elevated, the bottom-board will be found covered with pieces of bee-bread, &c. mixed with the *excrement of the worms* which looks almost exactly like fine grains of powder. As the bees in Spring, clean out their combs, and prepare the cells for the reception of brood, their bottom-board will often be so covered with parings of comb and with small pieces of bee-bread, that the hive may appear to be in danger of being destroyed by the worms. If, however, none of the *black* excrement is perceived, the refuse on the bottom-board, like the shavings in a carpenter's shop, are proofs of industry and not the signs of approaching ruin. It is highly important, however, to keep the bottom-boards clean, and if a piece of zinc be slipt in, (or even an old newspaper,) by removing and cleansing it from time to time, the bees will be greatly assisted in their operations. As soon as the hive is well filled with bees, this need no longer be done.

Even the most careful and experienced Apiarian will find, too often, that although he is perfectly well aware of the plague that is reigning within, his knowledge can be turned to no good account, the interior of the hive being almost as inaccessible as the interior of the human body. The way in which I manage, in such cases, is as follows.

Having ascertained, in the Spring, as soon as the bees begin to fly out, that a colony although feeble, has a fertile queen, I take the precaution at once to give it the strength which is indispensable, not merely to its safety, but to its ability for any kind of successful labor.

As a certain number of bees are needed in a hive, in order as well to warm and hatch the thousands of eggs which a healthy queen can lay, as to feed and properly develop the larvæ after they are hatched, I know that a feeble colony must remain feeble for a long time, unless they can at once be supplied with a considerable accession of numbers. Even if there were no moths in existence to trouble such a hive, it would not be able to rear a large number of bees, until after the best of the honey-harvest had passed away: and then it would become powerful only that its increasing numbers might devour the food which the others had previously stored in the cells. If the small colony has a considerable number of bees, and is able to cover and warm at least one comb in addition to those containing brood which they already have, I take from one of my strong stocks, a frame containing some three or four thousand or more young bees, which are sealed over in their cells, and are just ready to emerge. These bees which require no food, and need nothing but warmth to develop them, will, in a few days, hatch in the new hive to which they are given, and thus the requisite number of workers, in the full vigor and energy of youth, will be furnished to the hive, and the discouraged queen, finding at once a suitable number of experienced nurses[23] to take charge of her eggs, deposits them in the proper cells, instead of simply extruding them, to be devoured by the bees. While bees often attack full grown strangers which are introduced into their hive, they never fail to receive gladly all the brood comb that we choose to give them. If they are sufficiently numerous, they will always cherish it, and in warm weather, they will protect it, even if it is laid against the outside of their hive! If the bees in the weak stock, are too much reduced in numbers, to be able to cover the brood comb taken from another hive, I give them this comb with all the old bees that are clustered upon it, and shut up the hive, after supplying them with water, until two or three days have passed away. By this time, most of the strange bees will have formed an inviolable attachment to their new home, and even if a portion of them should return to the parent hive, a large number of the maturing young will have hatched, to supply their desertion. A little sugar-water scented with peppermint, may be used to sprinkle the bees, at the time that the comb is introduced, although I have never yet found that they had the least disposition, to quarrel with each other. The original settlers are only too glad to receive such a valuable accession to their scanty numbers, and the expatriated bees are too-much confounded with their unexpected emigration, to feel any desire for making a disturbance. If a sufficient increase of numbers has not been furnished by one range of comb, the operation may, in the course of a few days, be repeated. Instead of leaving the colony to the discouraging feeling that they are in a large, empty and desolate house, a divider should be run down into the hive, and they should be confined to a space which they are able to warm and defend, and the rest of the hive, until they need its additional room, should be carefully shut up against all intruders. If this operation is judiciously performed, the bees will be powerful in numbers, long before the weather is warm enough to develop the bee-moth, and they will thus be most effectually protected from the hateful pest.

A very simple change in the organization of the bee-moth would have rendered it almost if not quite impossible to protect the bees from its ravages. If it had been so constituted as to require but a very small amount of heat for its full development, it would have become very numerous early in the Spring, and might then have easily entered the hives and deposited its eggs among the combs, without any let or hindrance; for at this season, not only do the bees at night maintain no guard at the entrance of their hive, but there are large portions of their comb bare of bees, and of course, entirely unprotected. How does every fact in the history of the bee, when properly investigated, point with unerring certainty to the power, wisdom and goodness, of Him who made it!

If there is reason to apprehend that the combs which are not occupied with brood, contain any of the eggs of the moth, these combs may be removed, and thoroughly smoked with the fumes of burning sulphur; and then, in a few days, after they have been exposed to the fresh air, they may be returned to the hive. I hope I may be pardoned for feeling not the slightest pity for the unfortunate progeny of the moth, thus unceremoniously destroyed.

Bees, as is well known to every experienced bee-keeper, frequently swarm so often as to expose themselves to great danger of being destroyed by the moth. After the departure of the after-swarms, the parent colony often contains too few bees to cover and protect their combs from the insidious attacks of their wily enemy. As a number of weeks must elapse before the brood of the young queen is mature, the colony, for a considerable time, at the season when the moths are very numerous, are constantly diminishing in numbers, and before they can begin to replenish the exhausted hive, the destroyer has made a fatal lodgment.

In my hives, such calamities are easily prevented. If artificial increase is relied upon for the multiplication of colonies, it can be so conducted as to give the moth next to no chance to fortify itself in the hive. No colony is ever allowed to have more room than it needs, or more combs than it can cover and protect; and the entrance to the hive may be contracted, if necessary, so that only a single bee can go in and out, at a time, and yet the bees will have, from the ventilators, as much air as they require.

If natural swarming is allowed, after-swarms may be prevented from issuing, by cutting out all the queen cells but one, soon after the first swarm leaves the hive; or if it is desired to have as fast an increase of stocks, as can possibly be obtained from natural swarming, then instead of leaving the combs in the parent hive to be attacked by the moth, a certain portion of them may be taken out, when swarming is over, and given to the second and third swarms, so as to aid in building them up into strong stocks.

But I have not yet spoken of the most fruitful cause of the desolating ravages of the bee-moth. If a colony has *lost its queen*, and this loss cannot be supplied, it must, as a matter of course, fall a sacrifice to the bee-moth: and I do not hesitate to assert that by far the larger proportion of colonies which are destroyed by it, are destroyed under precisely such circumstances! Let this be remembered by all who have any thing to do with bees, and let them understand that unless a remedy for the loss of the queen, can be provided, they must constantly expect to see some of their best colonies hopelessly ruined. The crafty moth, after all, is not so much to blame, as we are apt to imagine; for a colony, once deprived of its queen, and possessing no means of securing another, would certainly perish, even if never attacked by so deadly an enemy; just as the body of an animal, when deprived of life, will speedily go to decay, even if it is not, at once, devoured by ravenous swarms of filthy flies and worms.

In order to ascertain all the important points connected with the habits of the bee-moth, I have purposely deprived colonies in some of my observing hives, of their queen, and have thus reduced them to a state of despair, that I might closely watch all their proceedings. I have invariably found that in this state, they have made little or no resistance to the entrance of the bee-moth, but have allowed her to deposit her eggs, just where she pleased. The worms, after hatching, have always appeared to be even more at home than the poor dispirited bees themselves, and have grown and thrived, in the most luxurious manner. In some instances, these colonies, so far from losing all spirit to resent other intrusions, were positively the most vindictive set of bees in my whole Apiary. One especially, assaulted every body that came near it, and when reduced in numbers to a mere handful, seemed as ready for fight as ever.

How utterly useless then, for defending a queenless colony against the moth, are all the traps and other devices which have been, of late years, so much relied upon. If a single female gains admission, she will lay eggs enough to destroy in a short time, the strongest colony that ever existed, if once it has lost its queen, and has no means of procuring another. But not only do the bees of a hive which is hopelessly queenless, make little or no opposition to the entrance of the bee-moth, and to the ravages of the worms, but by their forlorn condition, they positively invite the attacks of their destroyers. The moth seems to have an instinctive knowledge of the condition of such a hive, and no art of man can ever keep her out. She will pass by other colonies to get at the queenless one, for she seems to know that there she will find all the conditions that are necessary to the proper development of her young. There are many mysteries in the insect world, which we have not yet solved; nor can we tell just how the moth arrives at so correct a knowledge of the condition of the queenless hives in the Apiary. That such hives, very seldom, maintain a guard about the entrance, is certain; and that they do not fill the air with the pleasant voice of happy industry, is equally certain; for even to our dull ears, the difference between the hum of the prosperous hive, and the unhappy note of the despairing one, is sufficiently obvious. May it not be even more obvious to the acute senses of the provident mother, seeking a proper place for the development of her young?

The unerring sagacity of the moth, closely resembles that peculiar instinct by which the vulture and other birds that prey upon carrion, are able to single out a diseased animal from the herd, which they follow with their dismal croakings, hovering over its head, or sitting in ill-omened flocks, on the surrounding trees, watching it as its life ebbs away, and stretching out their filthy and naked necks, and opening and snapping their blood-thirsty beaks that they may be all ready to tear out its eyes just glazing in death, and banquet upon its flesh still warm with the blood of life! Let any fatal accident befall an animal, and how soon will you see them, first from one quarter of the heavens, and then from another, speeding their eager flight to their destined prey, when only a short time before, not a single one could be seen or heard.

I have repeatedly seen powerful colonies speedily devoured by the worms, because of the loss of their queen, when they have stood, side by side with feeble colonies which being in possession of a queen, have been left untouched!

That the common hives furnish no available remedy for the loss of the queen, is well known: indeed, the owner cannot, in many cases, be sure that his bees are queenless, until their destruction is certain, while not infrequently, after keeping bees for many years, he does not even so much as believe that there is such a thing as a queen bee!

In the Chapter on the Loss of the Queen, I shall show in what way this loss can be ascertained, and ordinarily remedied, and thus the bees be protected from that calamity which more than all others, exposes them to destruction. When a colony has become hopelessly queenless, then moth or no moth, its destruction is absolutely certain. Even if the bees retained their wonted industry in gathering stores, and their usual energy in defending themselves against all their enemies, their ruin could only be delayed for a short time. In a few months, they would all die a natural death, and there being none to replace them, the hive would be utterly depopulated. Occasionally, such instances occur in which the bees have died, and large stores of honey have been found untouched in their hives. This, however, but seldom happens: for they rarely escape from the assaults of other colonies, even if after the death of their queen, they do not fall a prey to the bee-moth. A motherless hive is almost always assaulted by stronger stocks, which seem to have an instinctive knowledge of its orphanage, and hasten at once, to take possession of its spoils. (See Remarks on Robbing.) If it escape the Scylla of these pitiless plunderers, it is soon dashed upon a more merciless Charybdis, when the miscreant moths have ascertained its destitution. Every year, large numbers of hives are bereft of their queen, and every year, the most of such hives are either robbed by other bees, or sacked by the bee-moth, or first robbed, and afterwards sacked, while their owner imputes all the mischief that is done, to something else than the real cause. He might just as well imagine that the birds, or the carrion worms which are devouring his dead horse, were actually the primary cause of its untimely end. How often we see the same kind of mistake made by those who impute the decay of a tree, to the insects which are banqueting upon its withering foliage; when often these insects are there, because the disease of the tree has both furnished them with their proper aliment, and deprived the plant of the vigor necessary to enable it to resist their attack.

The bee keeper can easily gather from these remarks, the means upon which I most rely, to protect my colonies from the bee-moth. Knowing that strong stocks supplied with a fertile queen, are always able to take care of themselves, in almost any kind of hive, I am careful to keep them in the state which is practically found to be one of such security. If they are weak, they must be properly strengthened, and confined to only as much space as they can warm and defend: and if they are queenless, they must be supplied with the means of repairing their loss, or if that cannot be done, they should be at once broken up, (See Remarks on Queenlessness, and Union of Stocks,) and added to other stocks.

It cannot be too deeply impressed upon the mind of the bee-keeper, that a small colony ought always to be confined to a small space, if we wish the bees to work with the greatest energy, and to offer the stoutest resistance to their numerous enemies. Bees do most unquestionably, "abhor a vacuum," if it is one which they can neither fill, warm nor defend. Let the prudent bee-master only keep his stocks strong, and they will do more to defend themselves against all intruders, than he can possibly do for them, even if he spends his whole time in watching and assisting them.

It is hardly necessary, after the preceding remarks, to say much upon the various contrivances to which so many resort, as a safeguard against the bee-moth. The idea that gauze-wire doors, to be shut daily at dusk, and opened again at morning, can exclude the moth, will not weigh much with one who has seen them flying and seeking admission, especially in dull weather, long before the bees have given over their work for the day. Even if the moth could be excluded by such a contrivance, it would require, on the part of those who rely upon it, a regularity almost akin to that of the heavenly bodies in their courses; a regularity so systematic, in short, as either to be impossible, or likely to be attained but by very few.

An exceedingly ingenious contrivance, to say the least, to remedy the necessity for such close supervision, is that by which the movable doors of all the hives are governed by a long lever in the shape of a hen-roost, so that the hives may all be closed seasonably and regularly, by the crowing and cackling tribe, when they go to bed at night, and opened at once when they fly from their perch, to greet the merry morn. Alas! that so much ingenuity should be all in vain! Chickens are often sleepy, and wish to retire sometime before the bees feel that they have completed their full day's work, and some of them are so much opposed to early rising, either from ill-health, or downright laziness, that they sit moping on their roost, long after the cheerful sun has purpled the glowing East. Even if this device were perfectly successful, it could not save from ruin, a colony which has lost its queen. The truth is, that almost all the contrivances upon which we are instructed to rely, are just about equivalent to the lock carefully put upon the stable door, after the horse has been stolen; or to attempts to prevent corruption from fastening upon the body of an animal, after the breath of life has forever departed.

Are there then no precautions to which we may resort, except by using hives which give the control over every comb? Certainly there are, and I shall now describe them in such a manner as to aid all who find themselves annoyed by the inroads of the bee-moth.

Let the prudent bee-master be deeply impressed with the very great importance of destroying *early* in the season, the larvæ of the bee-moth. "Prevention is," at all times, "better than cure": a single pair of worms that are permitted to undergo their changes into the winged insect, may give birth to some hundreds which before the close of the season, may fill the Apiary with thousands of their kind. The destruction of a single worm early in the Spring, may thus be more efficacious than that of hundreds, at a later period. If the common hives are used, these worms must be sought for in their hiding places, under the edges of the hive; or the hive may be propped up, on the two ends, with strips of wood, about three eighths of an inch thick; and a piece of old woolen rag put between the bottom-board and the back of the hive. Into this warm hiding place, the full grown worm will retreat to spin its cocoon, and it may then be very easily caught and effectually dealt with. Hollow sticks, or split joints of cane may be set under the hives, so as to elevate them, or may be laid on the bottom-board, and if they have a few small openings through which the bees cannot enter, the worms will take possession of them, and may easily be destroyed. Only provide some hollow, inaccessible to the bees, but communicating with the hive and easily accessible to the worms when they want to spin, and to yourself when you want them, and if the bees are in good health, so that they will not permit the worms to spin among the combs, you can, with ease, entrap nearly all of them. If the hive has lost its queen, and the worms have gained possession of it, you can do nothing for it better than to break it up as soon as possible, unless you prefer to reserve it as a moth trap to devastate your whole Apiary.

I make use of blocks of a peculiar construction, in order both to entrap the worms, and to exclude the moth from my hives. The only place where the moth can enter, is just where the bees are going in and out, and this passage may be contracted so as to suit the size of the colony:

the very shape of it is such that if the moth attempts to force an entrance, she is obliged to travel over a space which is continually narrowing, and of course, is more and more easily defended by the bees. My traps are slightly elevated, so that the heat and odor of the hive pass under them, and come out through small openings into which the moth can enter, but which do not admit her into the hive. These openings, which are so much like the crevices between the common hives and their bottom-boards, the moth will enter, rather than attempt to force her way through the guards, and finding here the nibblings and parings of comb and bee-bread, in which her young can flourish, she deposits her eggs in a place where they may be reached and destroyed. All this is on the supposition that the hive has a healthy queen, and that the bees are confined to a space which they can warm and defend. If there are no guards and no resistance, or at best but a very feeble one, she will not rest in any outer chamber, but will penetrate to the very heart of the citadel, and there deposit her seeds of mischief. These same blocks have also grooves which communicate with the *interior* of the hives, and which appear to the prowling worm in search of a comfortable nest, just the very best possible place, so warm and snug and secure, in which to spin its web, and "bide its time." When the hand of the bee-master lights upon it, doubtless it has reason to feel that it has been caught in its own craftiness.

If asked how much will such contrivances help the careless bee-man, I answer, not one iota; nay, they will positively furnish him greater facilities for destroying his bees. Worms will spin and hatch, and moths will lay their eggs, under the blocks, and he will never remove them: thus instead of traps he will have most beautiful devices for giving more effectual aid and comfort to his enemies. Such persons, if they ever attempt to keep bees on my plans, should use only my smooth blocks, which will enable them to control, at will, the size of the entrance to the hives, and which are exceedingly important in aiding the bees to defend themselves against moths and robbers, and all enemies which seek admission to their castle.

Let me, however, strongly advise the thoroughly and incorrigibly careless, to have nothing to do with bees, either on my plan of management, or any other; for they will find their time and money almost certainly thrown away; unless their mishaps open their eyes to the secret of their failure in other things, as well as in bee-keeping.

If I find that the worms, by any means have got the upper hand in one of my hives, I take out the combs, shake off the bees, route out the worms and restore the combs again to the bees: if there is reason to fear that they contain eggs and small worms, I smoke them thoroughly with sulphur, and air them well before they are returned. Such operations, however, will very seldom be required. Shallow vessels containing sweetened water, placed on the hives after sunset, will often entrap many of the moths. Pans of milk are recommended by some as useful for this purpose. So fond are the moths of something sweet, that I have caught them *sticking fast* to pieces of moist sugar-candy.

I cannot deny myself the pleasure of making an extract from an article[24] from the pen of that accomplished scholar, and well-known enthusiast in bee-culture, Henry K. Oliver, Esq. "We add a few words respecting the enemies of bees. The mouse, the toad, the ant, the stouter spiders, the wasp, the death-head moth, (Sphinx atropos,) and all the varieties of gallinaceous birds, have, each and all, "a sweet tooth," and like, very well, a dinner of raw bee. But the ravages of all these are but a baby bite to the destruction caused by the bee-moth, (Tinea mellonella.) These nimble-footed little mischievous vermin may be seen, on any evening, from early May to October, fluttering about the apiary, or running about the hives, at a speed to outstrip the swiftest bee, and endeavoring to effect an entrance into the door way, for it is within the hive that their instinct teaches them they must deposit their eggs. You can hardly find them by day, for they are cunning and secrete themselves. "They love darkness rather than light, because their deeds are evil." They are a paltry looking, insignificant little grey-haired pestilent race of wax-and-honey-eating and bee-destroying rascals, that have baffled all contrivances that ingenuity has devised to conquer or destroy them."

"Your committee would be very glad indeed to be able to suggest any effectual means, by which to assist the honey-bee and its friends, against the inroads of this, its bitterest and most successful foe, whose desolating ravages are more lamented and more despondingly referred to, than those of any other enemy. Various contrivances have been announced, but none have proved efficacious to any full extent, and we are compelled to say that there really is no security, except in a very full, healthy and vigorous stock of bees, and in a very close and well made hive, the door of which is of such dimensions of length and height, that the nightly guards can effectually protect it. Not too long a door, nor too high. If too long, the bees cannot easily guard it, and if too high, the moth will get in over the heads of the guards. If the guards catch one of them, her life is not worth insuring. But if the moths, in any numbers, effect a lodgment in the hive, then the hive is not worth insuring. They immediately commence laying their eggs, from which comes, in a few days, a brownish white caterpillar, which encloses itself, all but its head, in a silken cocoon. This head, covered with an impenetrable coat of scaly mail, which bids defiance to the bees, is thrust forward, just outside of the silken enclosure, and the gluttonous pest eats all before it, wax, pollen, and exuviæ, until ruin to the stock is inevitable. As says the Prophet Joel, speaking of the ravages of the locust, "the land is as the garden of Eden before them, and behind them a desolate wilderness." Look out, brethren, bee lovers, and have your hives of the best unshaky, unknotty stock, with close fitting joints, and well covered with three or four coats of paint. He who shall be successful in devising the means of ridding the bee world of this destructive and merciless pest, will richly deserve to be crowned "King Bee," in perpetuity, to be entitled to a never-fading wreath of budding honey flowers, from sweetly breathing fields, all murmuring with bees, to be privileged to use, during his natural life, "night tapers from their waxen thighs," best wax candles, (two to the pound!) to have an annual offering from every bee-master, of ten pounds each, of very best virgin honey, and to a body guard, for protection against all foes, of thrice ten thousand workers, all armed and equipped, as Nature's law directs. Who shall have these high honors?"

It might seem highly presumptuous for me, at this early date, to lay claim to them, but I beg leave to enroll myself among the list of honorable candidates, and I cheerfully submit my pretensions to the suffrages of all intelligent keepers of bees.

In the chapter on Requisites, I have spoken of the ravages of the mouse, and have there described the way in which my hives are guarded against its intrusion. That some kinds of birds are fond of bees, every Apiarian knows, to his cost; still, I cannot advise that any should, on this account, be destroyed. It has been stated to me, by an intelligent observer, that the King-bird, which devours them by scores, confines himself always, in the season of drones, to those fat and lazy gentlemen of leisure. I fear however, that this, as the children say, "is too good news to be true," and that not only the industrious portion of the busy community fall a prey to his fatal snap, but that the luxurious gourmand can distinguish perfectly well, between an empty bee in search of food, and one which is returning full laden to its fragrant home, and whose honey-bag sweetens the delicious tit-bit, as the crushed unfortunate, all ready sugared, glides daintily down his voracious maw! Still, I have never yet been willing to destroy a bird, because of its fondness for bees; and I advise all lovers of bees to have nothing to do with such foolish practices. Unless we can check among our people, the stupid, as well as inhuman custom of destroying so wantonly, on any pretence, and often on none at all, the insectivorous birds, we shall soon, not only be deprived of their aerial melody, among the leafy branches, but shall lament over the ever increasing horde of destructive insects, which ravage our fields and desolate our orchards, and from whose successful inroads, nothing but the birds can ever protect us. Think of it, ye who can enjoy no music made by these winged choristers of the skies, except that of their agonizing screams, as they fall before your well-aimed weapons, and flutter out their innocent lives before your heartless gaze! Drive away as fast and as far as you please, from your cruel premises, all the little birds that you cannot destroy, and then find, if you can, those who will sympathize with you, when the caterpillars weave their destroying webs over your leafless trees, and insects of all kinds riot in glee, upon your blasted harvests! I hope that such a healthy public opinion will soon prevail, that the man or boy who is armed with a gun to shoot the little birds, will be scouted from all humane and civilized society, and if he should be caught about such contemptible business, will be too much ashamed even to look an honest man in the face. I shall close what I have to say about the birds, with the following beautiful translation of an old Greek poet's address to the swallow.

"Attic maiden, honey fed, Chirping warbler, bear'st away, Thou the busy buzzing bee, To thy callow brood a prey? Warbler, thou a warbler seize? Winged, one with lovely wings? Guest thyself, by Summer brought, Yellow guest whom Summer brings? Wilt not quickly let it drop? 'Tis not fair, indeed 'tis wrong, That the ceaseless warbler should Die by mouth of ceaseless song." *Merivale's Translation.*

I have not the space to speak at length of the other enemies of the honied race: nor indeed is it at all necessary. If the Apiarian only succeeds in keeping his stocks strong, they will be their own best protectors, and if he does not succeed in this, they would be of little value, even if they had no enemies ever vigilant, to watch for their halting. Nations which are both rich and feeble, invite attack, as well as unfit themselves for vigorous resistance. Just so with the commonwealth of bees. Unless amply guarded by thousands ready to die in its defence, it is ever liable to fall a prey to some one of its many enemies, which are all agreed in this one opinion, at least, that stolen honey is much more sweet than the slow accumulations of patient industry.

In the Chapters on Protection and Ventilation, I have spoken of the fatal effects of dysentery. This disease can always be prevented by proper caution on the part of the bee-keeper. Let him be careful not to feed his bees, late in the season, on liquid honey, (see Chapter on Feeding,) and let him keep them in dry and thoroughly protected hives. If his situation is at all damp, and there is danger that water will settle under his Protector, let him build it entirely *above ground*; otherwise it may be as bad as a damp cellar, and incomparably worse than nothing at all.

There is one disease, called by the Germans, "foul brood," of which I know nothing, by my own observation, but which is, of all others, the most fatal in its effects. The brood appear to die in the cells, after they are sealed over by the bees, and the stench from their decaying bodies infects the hive, and seems to paralyze the bees. This disease is, in two instances, attributed by Dzierzon, to feeding bees on "American Honey," or, as we call it, Southern Honey, which is brought from Cuba, and other West India Islands. That such honey is not ordinarily poisonous, is well known: probably that used by him, was taken from diseased colonies. It is well known that if any honey or combs are taken from a hive in which this pestilence is raging, it will most surely infect the colonies to which they may be given. No foreign honey ought therefore to be extensively used, until its quality has been thoroughly tested. The extreme violence of this disease may be inferred from the fact, that Dzierzon in one season, lost by it, between four and five hundred colonies! As at present advised, if my colonies were attacked by it, I should burn up the bees, combs, honey, frames, and all, from every diseased hive; and then thoroughly scald and smoke with sulphur, all such hives, and replenish them with bees from a healthy stock.

There is a peculiar kind of dysentery which does not seem to affect a whole colony, but confines its ravages to a small number of the bees. In the early stages of this disease, those attacked are excessively irritable, and will attempt to sting any one who comes near the hives. If dissected, their stomachs are found to be already discolored by the disease. In the latter stages of this complaint, they not only lose all their irascibility, but seem very stupid, and may often be seen crawling upon the ground unable to fly. Their abdomens are now unnaturally swollen, and of a much lighter color than usual, owing to their being filled with a yellow matter exceedingly offensive to the smell. I have not yet ascertained the cause of this disease.

CHAPTER XIII. LOSS OF THE QUEEN.

That the queen of a hive is often lost, and that the ruin of the whole colony soon follows, unless such a loss is seasonably remedied, are facts which ought to be well known to every observing bee-keeper.

Some queens appear to die of old age or disease, and at a time when there are no worker-eggs, or larvæ of a suitable age, to enable the bees to supply their loss. It is evident, however, that no very large proportion of the queens which perish, are lost under such circumstances. Either the bees are aware of the approaching end of their aged mother, and take seasonable precautions to rear a successor; or else she dies very suddenly, so as to leave behind her, brood of a suitable age. It is seldom that a queen in a hive that is strong in numbers and stores, dies either at a period of the year when there is no brood from which another can be reared, or when there are no drones to impregnate the one reared in her place. In speaking of the age of bees, it has already been stated that queens commonly die in their fourth year, while none of the workers live to be a year old. Not only is the queen much longer lived than the other bees, but she seems to be possessed of greater tenacity of life, so that when any disease overtakes the colony, she is usually among the last to perish. By a most admirable provision, their death ordinarily takes place under circumstances the most favorable to their bereaved family. If it were otherwise, the number of colonies which would annually perish, would be very much greater than it now is; for as a number of superannuated queens must die every year, many, or even most of them might die at a season when their loss would necessarily involve the ruin of their whole colony. In non-swarming hives, I have found cells in which queens were reared, not to lead out a new swarm, but to supply the place of the old one which had died in the hive. There are a few well authenticated instances, in which a young queen has been matured before the death of the old one, but after she had become quite aged and infirm. Still, there are cases where old queens die, either so suddenly as to leave no young brood behind them, or at a season when there are no drones to impregnate the young queens.

That queens occasionally live to such an age as to become incapable of laying worker eggs, is now a well established fact. The seminal reservoir sometimes becomes exhausted, before the queen dies of old age, and as it is never replenished, she can only lay unimpregnated eggs, or such as produce drones instead of workers. This is an additional confirmation of the theory first propounded by Dzierzon. I am indebted to Mr. Wagner for the following facts. "In the Bienenzeitung, for August, 1852, Count Stosch gives us the case of a colony examined by himself, with the aid of an experienced Apiarian, on the 14th of April, previous. The worker-brood was then found to be healthy. In May following, the bees worked industriously, and built new comb. Soon afterwards they ceased to build, and appeared dispirited; and when, in the beginning of June, he examined the colony again, he found plenty of drone brood in worker cells! The queen appeared weak and languid. He confined her in a queen cage, and left her in the hive. The bees clustered around the cage; but next morning the queen was found to be dead. Here we seem to have the commencement, progress and termination of super-annuation, all in the space of five or six weeks."

In the Spring of the year, as soon as the bees begin to fly, if their motions are carefully watched, the Apiarian may even in the common hives, generally ascertain from their actions, whether they are in possession of a fertile queen. If they are seen to bring in bee-bread with great eagerness, it follows, as a matter of course, that they have brood, and are anxious to obtain fresh food for its nourishment. If any hive does not industriously gather pollen, or accept the rye flour upon which the others are feasting, then there is an almost absolute certainty either that it has not a queen, or that she is not fertile, or that the hive is seriously infested with worms, or that it is on the very verge of starvation. An experienced eye will decide upon the queenlessness, (to use the German term,) of a hive, from the restless appearance of the bees. At this period of the year when they first realize the magnitude of their loss, and before they have become in a manner either reconciled to it, or indifferent to their fate, they roam in an inquiring manner, in and out of the hive, and over its outside as well as inside, and plainly manifest that something calamitous has befallen them. Often those that return from the fields, instead of entering the hive with that dispatchful haste so characteristic of a bee returning well stored to a prosperous home, linger about the entrance with an idle and very dissatisfied appearance, and the colony is restless, long after the other stocks are quiet. Their home, like that of the man who is cursed rather than blessed in his domestic relations, is a melancholy place: and they only enter it with reluctant and slow-moving steps!

If I could address a friendly word of advice to every married woman, I would say, "Do all that you can to make your husband's home a place of attraction. When absent from it, let his heart glow at the very thought of returning to its dear enjoyments; and let his countenance involuntarily put on a more cheerful look, and his joy-quickened steps proclaim, as he is approaching, that he feels in his "heart of hearts," that "there is no place like home." Let her whom he has chosen as a wife and companion, be the happy and honored Queen in his cheerful habitation: let her be the center and soul about which his best affections shall ever revolve. I know that there are brutes in the guise of men, upon whom all the winning attractions of a prudent, virtuous wife, make little or no impression. Alas that it should be so! but who can tell how many, even of the most hopeless cases, have been saved for two worlds, by a union with a virtuous woman, in whose "tongue was the law of kindness," and of whom it could be said, "the heart of her husband doth safely trust in her," for "she will do him good and not evil, all the days of her life."

Said a man of large experience, "I scarcely know a woman who has an intemperate husband, who did not either marry a man whose habits were already bad, or who did not drive her husband to evil courses, (often when such a calamitous result was the furthest possible from her thoughts or wishes,) by making him feel that he had no happy home." Think of it, ye who find that home is not full of dear delights, as well to yourselves, as to your affectionate husbands! Try how much virtue there may be in winning words and happy smiles, and the cheerful discharge of household duties, and prove the utmost possible efficacy of love and faith and prayer, before those words of fearful agony are extorted from your despairing lips,

"Anywhere, anywhere Out of the world;"

when amid tears and sighs of inexpressible agony, you settle down into the heart-breaking conviction that you can have no home until you have passed into that habitation not fashioned by human hands, or inhabited by human hearts!

Is there any husband who can resist all the sweet attractions of a lovely wife? who does not set a priceless value upon the very gem of his life?

"If such there be, go mark him well; High though his titles, proud his fame, Boundless his wealth as wish can claim, The wretch, concentered all in self, Living, shall forfeit fair renown, And doubly dying, shall go down To the vile dust from whence he sprung Unwept, unhonored, and unsung."—*Scott.*

I trust my readers, remembering my profession, will pardon this long digression to which I felt myself irresistibly impelled.

When the bees commence their work in the Spring, they give, as previously stated, reliable evidence either that all is well, or that ruin lurks within. In the common hives however, it is not always easy to decide upon their real condition. The queenless ones do not, in all cases, disclose their misfortune, any more than all unhappy husbands or wives see fit to proclaim the full extent of their domestic wretchedness: there is a vast amount of *seeming* even in the little world of the bee-hive. One great advantage in my mode of construction is that I am never obliged to leave anything to vague conjecture; but I can, in a few moments, open the interior, and know precisely what is the real condition of the bees.

On one occasion I found that a colony which had been queenless for a considerable time, utterly refused to raise another, and devoured all the eggs which were given to them for that purpose! This colony was afterwards supplied with an unimpregnated queen, but they refused to accept of her, and attempted at once to smother her to death. I then gave them a fertile queen, but she met with no better treatment. Facts of a similar kind have been noticed, by other observers: thus it seems that bees may not only become reconciled, as it were, to living without a mother, but may pass into such an unnatural state as not only to decline to provide themselves with another, but actually to refuse to accept of one by whose agency they might be rescued from impending ruin! Before expressing too much astonishment at such foolish conduct, let us seriously inquire if it has not often an exact parallel in our obstinate rejection of the provisions which God has made in the Gospel for our moral and religious welfare.

If a colony which refuses to rear another queen, has a range of comb given to it containing maturing brood, these poor motherless innocents, as soon as they are able to work, perceive their loss, and will proceed at once, if they have the means, to supply it! They have not yet grown so hardened by habit to unnatural and ruinous courses, as not to feel that something absolutely indispensable to their safety is wanting in their hive.

A word to the young who may read this treatise. Although enjoined to "remember your Creator in the days of your youth," you are constantly tempted to neglect your religious duties, and to procrastinate their performance until some more convenient season. Like the old bees in a hive without a queen, that seek only their present enjoyment, forgetful of the ruin which must surely overtake them, so you may find that when manhood and old age arrive, you will have even less disposition to love and serve the Lord than you now have. The fetters which bind you to sinful habits will have strengthened with years until you find both the inclination and ability to break them continually decreasing.

In the Spring, as soon as the weather becomes sufficiently pleasant, I carefully examine all the hives which do not present the most unmistakable evidences of health and vigor. If a queen is wanting, I at once, if the colony is small, break it up, and add the bees to another stock. If however, the colony should be very large, I sometimes join to it one of my small stocks which has a healthy queen. It may be asked why not supply the queenless stock with the means of raising another? Simply because there would be no drones to impregnate her, in season; and the whole operation would therefore result in an entire failure. Why not endeavor then to preserve it, until the season for drones approaches, and then give it a queen? Because it is in danger of being robbed or destroyed by the moth, while the bees, if added to another stock, can do me far more service than they could, if left to idleness in their old hive. It must be remembered that I am not like the bee-keepers on the old plan, extremely anxious to save every colony, however feeble: as I can, at the proper season, form as many as I want, and with far less trouble and expense than are required to make anything out of such discouraged stocks.

If any of my colonies are found to be feeble in the Spring, but yet in possession of a healthy queen, I help them to combs containing maturing brood, in the manner already described. In short, I ascertain, at the opening of the season, the exact condition of all my stock, and apply such remedies as I find to be needed, giving to some, maturing brood, to others honey, and breaking up all whose condition appears to admit of no remedy. If however, the bees have not been multiplied too rapidly, and proper care was taken to winter none but strong stocks, they will need but little assistance in the Spring; and nearly all of them will show indubitable signs of health and vigor.

I strongly recommend every prudent bee-keeper who uses my hives, to give them all a most thorough over-hauling and cleansing, soon after the bees begin to work in the Spring. The bees of any stock may, with their combs, &c., all be transferred, in a few minutes, to a clean hive; and their hive, after being thoroughly cleansed, may be used for another transferred stock; and in this way, with one spare hive, the bees may all be lodged in habitations from which every speck of dirt has been removed. They will thus have hives which can by no possibility, harbor any of the eggs, or larvæ of the moth, and which may be made perfectly free from the least smell of must or mould or anything offensive to the delicate senses of the bees. In making this thorough cleansing of all the hives, the Apiarian will necessarily gain an exact knowledge of the true condition of each stock, and will know which have spare honey, and which require food: in short, which are in need of help in any respect, and which have the requisite strength to lend a helping hand to others. If any hive needs repairing, it may be put into perfect order, before it is used again. Hives managed in this fashion, if the roofs and outside covers are occasionally painted anew, will last for generations, and will be found, on the score of cheapness, preferable, in the long run, to any other kind. But I ought to beg pardon of the Genius of American cheapness, who so kindly presides over the making of most of our manufactures, and under whose shrewd tuition we are fast beginning to believe that cheapness in the first cost of an article, is the main point to which our attention should be directed!

Let us to be sure, save all that we can in the cost of construction, by the greatest economy in the use of materials; let us compel every minute to yield the greatest possible practical result, by the employment of the most skillful workmen and the most ingenious machinery; but do let us learn that slighting an article, so as to get up a mere sham, having all the appearance of reality, with none of the substance, is the poorest possible kind of pretended economy; to say nothing of the tendency of such a system, to encourage in all the pursuits of life, the narrow and selfish policy of doing nothing thoroughly, but everything with reference to mere outside show, or the urgent necessities of the present moment.

We have yet to describe under what circumstances, by far the larger proportion of hives, become queenless. After the first swarm has gone out with the old mother, then both the parent stock and all the subsequent swarms, will have each a young queen which must always leave the hive in order to be impregnated. It sometimes happens that the wings of the young female are, from her birth, so imperfect that she either refuses to sally out, or is unable to return to the hive, if she ventures abroad. In either case, the old stock must, if left to its own resources, speedily perish.

Queens, in their contests with each other, are sometimes so much crippled as to unfit them for flight, and sometimes they are disabled by the rude treatment of the bees, who insist on driving them away from the royal cells. The great majority, however, of queens which are lost, perish when they leave the hive in search of the drones. Their *extra size* and *slower flight* make them a most tempting prey to the birds, ever on the watch in the vicinity of the hives; and many in this way, perish. Others are destroyed by sudden gusts of winds, which dash them against some hard object, or blow them into the water; for queens are by no means, exempt from the misfortunes common to the humblest of their race. Very frequently, in spite of all their caution in noticing the position and appearance of their habitation, before they left it, they make a fatal mistake on their return, and are imprisoned and destroyed as they attempt to enter the wrong hive. The precautions which should be used, to prevent such a calamity, have been already described. If these are neglected, those who build their hives of uniform size and appearance, will find themselves losing many more queens than the person who uses the old-fashioned boxes, hardly any two of which look just alike.

The bees seem to me, to have, as it were, an instinctive perception of the dangers which await their new queen when she makes her excursion in search of the drones, and often gather around her, and confine her, as though they could not bear to have her leave! I have repeatedly noticed them doing this, although I cannot affirm with positive certainty, why they do it. They are usually excessively agitated when the queen leaves, and often exhibit all the appearance of swarming. If the queen of an old stock is lost in this way, her colony will gradually dwindle away. If the queen of an after-swarm fails to return, the bees very speedily come to nothing, if they remain in the hive; as a general rule, however, they soon leave and attempt to add themselves to other colonies.

It would be highly interesting to ascertain in what way the bees become informed of the loss of their queen. When she is taken from them under such circumstances as to excite the whole colony, then we can easily see how they find out that she is gone; for when greatly excited, they always seek first to assure themselves of her safety; just as a tender mother in time of danger forgets herself in her anxiety for her helpless children! If however, the queen is carefully removed, so that the colony is not disturbed, it is sometimes a day, or even more, before they realize their loss. How do they first become aware of it? Perhaps some dutiful bee feels that it is a long time since it has seen its mother, and anxious to embrace her, makes diligent search for her through the hive! The intelligence that she cannot anywhere be found, is soon noised abroad, and the whole community are at once alarmed. At such times, instead of calmly conversing by merely touching each other's antennæ, they may be seen violently striking as it were, their antennæ together, and by the most impassioned demonstrations manifesting their agony and despair. I once removed a queen in such a manner as to cause the bees to take wing and fill the air in search of her. She was returned in a few minutes, and yet, on examining the colony, two days after, I found that they had actually commenced the building of royal cells, in order to raise another! The queen was unhurt and the cells were not tenanted. Was this work begun by some that refused for a long time to believe the others, when told that she was safe? Or was it begun from the apprehension that she might again be removed?

Every colony which has a new queen, should be watched, in order that the Apiarian may be seasonably apprised of her loss. The restless conduct of the bees, on the evening of the day that she fails to return, will at once inform the experienced bee-master of the accident which has befallen his hive. If the bees cannot be supplied with another queen, or with the means of raising one, if an old swarm it must be broken up, and the bees added to another stock; if a new swarm it must always be broken up, unless it can be supplied with a queen nearly mature, or else they will build combs unfit for the rearing of workers. By the use of my movable comb hives, all these operations can be easily performed. If any hives have lost their young queen, they may be supplied, either with the means of raising another, or with sealed queens from other hives, or, (if the plan is found to answer,) with mature ones from the "Nursery."

As a matter of precaution, I generally give to all my stocks that are raising young queens, or which have unimpregnated ones, a range of comb containing brood and eggs, so that they may, in case of any accident to their queen, proceed at once, to supply their loss. In this way, I prevent them from being so dissatisfied as to leave the hive.

About a week after the young queens have hatched, I examine all the hives which contain them, lifting out usually, some of the largest combs, and those which ought to contain brood. If I find a comb which has eggs or larvæ, I am satisfied that they have a fertile queen, and shut up the hive; unless I wish to find her, in order to deprive her of her wings, I can thus often satisfy myself in one or two minutes. If no brood is found, I suspect that the queen has been lost, or that she has some defect which has prevented her from leaving the hive. If the brood-comb which I put into the hive, contains any newly-formed royal cells, I *know*, without any further examination, that the queen has been lost. If the weather has been unfavorable, or the colony is quite weak, the young queen is sometimes not impregnated as early as usual, and an allowance of a few days must be made on this account. If the weather is favorable, and the colony a good one, the queen usually leaves, the day after she finds herself mistress of a family. In about two days more, she begins to lay her eggs. By waiting about a week before the examination is made, ample allowance, in most cases, is made.

Early in the month of September, I examine carefully all my hives, so as to see that in every respect, they are in suitable condition for wintering. If any need feeding, (See Chapter on Feeding,) they are fed at this time. If any have more vacant room than they ought to have, I partition off that part of the hive which they do not need. I always expect to find some brood in every healthy hive at this time, and if in any hive I find none, and ascertain that it is queenless, I either at once break it up, or if it is strong in numbers supply it with a queen, by adding to it some feebler stock. If bees, however, are properly attended to, at the season when their young queens are impregnated, it will be a very rare occurrence to find a queenless colony in the Fall.

The practical bee-keeper without further directions, will readily perceive how any operation, which in the common hives, is performed with difficulty, if it can be performed at all, is reduced to simplicity and certainty, by the control of the combs. If however, bee-keepers will be negligent and ignorant, no hive can possible make them very successful. If they belong to the fraternity of "no eyes," who have kept bees all their lives, and do not know that there is a queen, they will probably derive no special pleasure from being compelled to believe what they have always derided as humbug or book-knowledge; although I have seen some bee-keepers very intelligent on most matters, who never seem to have learned the first rudiments in the natural history of the bee. Those who cannot, or will not learn for themselves, or who have not the leisure or disposition to manage their own bees, may yet with my hives, entrust their care to suitable persons who may, at the proper time, attend to all their wants. Practical gardeners may find the management of bees for their employers, to be quite a lucrative part of their profession. With but little extra labor and with great certainty, they may, from time to time, do all that the prosperity of the bees require; carefully over-hauling them in the Spring, making new colonies, at the suitable period, if any are wanted, giving them their surplus honey receptacles, and removing them when full; and on the approach of Winter, putting all the colonies into proper condition, to resist its rigors. The business of the practical Apiarian, and that of the Gardener, seem very naturally to go together, and one great advantage of my hive and mode of management is the ease with which they may be successfully united.

Some Apiarians after all that has been said, may still have doubts whether the young queens leave the hive for impregnation; or may think that the old ones occasionally leave, even when they do not go out to lead a swarm. Such persons may, if they choose, easily convince themselves by the following experiments of the accuracy of my statements. About a week after hiving a second swarm, or after the birth of a young queen in a hive, and after she has begun to lay eggs, open the hive and remove her: carry her a few rods in front of the Apiary, and let her fly; she will at once enter her own hive and thus show that she has previously left it. If, however, an old queen is removed a short time after hiving the swarm, she will not be able to distinguish her own hive from any other, and will thus show that she has not left it, since the swarm was hived. If this experiment is performed upon an old queen, in a hive in which she was put the year before, when unimpregnated, the same result will follow; for as she never left it after that event, she will have lost all recollection of its relative position in the Apiary. The first of these experiments has been suggested by Dzierzon.

CHAPTER XIV. UNION OF STOCKS. TRANSFERRING BEES FROM THE COMMON HIVE. STARTING AN APIARY.

Frequent allusions have been made to the importance, for various reasons, of breaking up stocks and uniting them to other families in the Apiary. Colonies which in the early Spring, are found to be queenless, ought at once to be managed in this way, for even if not speedily destroyed by their enemies, they are only consumers of the stores which they gathered in their happier days. The same treatment should also be extended to all that in the Fall, are found to be in a similar condition.

As small colonies, even though possessed of a healthy queen, are never able to winter as advantageously as large ones, the bees from several such colonies ought to be put together, to enable them by keeping up the necessary supply of heat, to survive the Winter on a smaller supply of food. A certain quantity of animal heat must be maintained by bees, in order to live at all, and if their numbers are too small, they can only keep it up, by eating more than they would otherwise require. A small swarm will thus not infrequently, consume as much honey as one containing two or three times as many bees. These are facts which have been most thoroughly tested on a very large scale. If a hundred persons are required to occupy, with comfort, a church that is capable of accommodating a thousand, as much fuel or even more will be required, to warm the small number as the large one.

If the stocks which are to be wintered, are in the common hives, the condemned ones must be drummed out of their old encampment, sprinkled with sugar-water scented with peppermint, or some other pleasant odor, and added to the others, The colonies which are to be united ought if possible, to stand side by side, some time before this process is attempted. This can almost always be effected by a little management, for while it would not be safe to move a colony all at once, even a few yards to the right or left of the line of flight in which the bees sally out to the fields, (especially if other hives are near,) they may be moved a slight distance one day, and a little more the next, and so on, until we have them at last in the desired place.

As persons may sometimes be obliged to move their Apiaries, during the working season, I will here describe the way by which I was able to accomplish such a removal, so as to benefit, instead of injuring my bees. Selecting a pleasant day, I moved, early in the morning, a portion of my very best stocks. A considerable number of bees from these colonies, returned in the course of the day to the familiar spot; after flying about for some time, in search of their hives, (if the weather had been chilly many of them would have perished,) they at length entered those standing next to their old homes. More of the strongest were removed, on the next pleasant day: and this process was repeated, until at last only one hive was left in the old Apiary. This was then removed, and only a few bees returned to the old spot. I thus lost no more bees, in moving a number of hives, than I should have lost in moving one: and I conducted the process in such a way, as to strengthen some of my feeble stocks, instead of very seriously diminishing their scanty numbers. I have known the most serious losses to result from the removal of an Apiary, conducted in the manner in which a change of location is usually made.

The process of uniting colonies in my hive, is exceedingly simple. The combs may, after the two colonies are sprinkled, be at once lifted out from the one which is to be broken up, and put with all the bees upon them, directly into the other hive. If the Apiarian judges it best to save any of his very small colonies, he can confine them to one half or one third of the central part of the hive, and fill the two empty ends with straw, shavings, or any good non-conductor. Any one of my frames, can, in a few minutes, by having tacked to it a thin piece of board or paste-board, or even an old newspaper, be fashioned into a divider, which will answer all practical purposes, and if it is stuffed with cotton waste, &c., it will keep the bees uncommonly warm. If a *very* small colony is to be preserved over Winter, the queen must be confined, in the Fall, in a queen cage, to prevent the colony from deserting the hive.

I shall now show how the bee-keeper who wishes only to keep a given number of stocks, may do so, and yet secure from that number the largest quantity of surplus honey.

If his bees are kept in non-swarming hives, he may undoubtedly, reap a bounteous harvest from the avails of their industry. I do not however, recommend this mode of bee-keeping as the best: still there are many so situated that it may be much the best for them. Such persons, by using my hives, can pursue the non-swarming plan to the best advantage. They can by taking off the wings of their queens, be sure that their colonies will not suddenly leave them; a casualty to which all other non-swarming hives are sometimes liable; and by taking away the honey in small quantities, they will always give the bees plenty of spare room for storage, and yet avoid discouraging them, as is so often done when large boxes are taken from them. (See Chapter on Honey.)

By removing from time to time, the old queens, the colonies can all be kept in possession of queens, at the height of their fertility, and in this way a very serious objection to the non-swarming, or as it is frequently called, the storifying system, may be avoided. If at any time, new colonies are wanted, they may be made in the manner already described. In districts where the honey harvest is of very short continuance, the non-swarming plan may be found to yield the largest quantity of honey, and in case the season should prove unfavorable for the gathering of honey, it will usually secure the largest returns from a given number of stocks. I therefore prefer to keep a considerable number of my colonies, on the storifying plan, and am confident of securing from them, a good yield of honey, even in the most unfavorable seasons. If bee-keepers will pursue the same system, they will not only be on the safe side, but will be able to determine which method it will be best for them to adopt, in order to make the most from their bees. As a general rule, the Apiarian who increases the number of his colonies, one third in a season, making one very powerful swarm from two, will have more surplus honey from the three, than he could have obtained from the two, to say nothing of the value of his new swarms. If, at the approach of Winter, he wishes to reduce his stocks down to the Spring number, he may unite them in the manner described, appropriating all the good honey of those which he breaks up, and saving all their empty comb for the new colonies of the next season. The bees in the doubled stock will winter most admirably; will consume but little honey, in proportion to their numbers, and will be in most excellent condition when the Spring opens. It must not, however, be forgotten, that although they eat comparatively little in the Winter, they must be well supplied in the Spring; as they will then have a very large number of mouths to feed, to say nothing of the thousands of young bees bred in the hive. If any old-fashioned bee-keeper wishes, he can thus pursue the old plan, with only this modification; that he preserves the lives of the bees in the hives which he wishes to take up; secures his honey without any fumes of sulphur, and saves the empty comb to make it worth nearly ten times as much to himself, as it would be, if melted into wax. Let no humane bee-keeper ever feel that there is the slightest necessity for so managing his bees as to make the comparison of Shakespeare always apposite:

"When like the Bee, tolling from every flower The virtuous sweets; Our thighs packed with wax, our mouths, with honey, We bring it to the hive; and like the bees, Are murdered for our pains."

While I am an advocate for breaking up all stocks which cannot be wintered advantageously, I never advise that a single bee should be killed. Self interest and Christianity alike forbid the unnecessary sacrifice.

TRANSFERRING BEES FROM THE COMMON HIVE TO THE MOVABLE COMB HIVE.

The construction of my hive is such, as to permit me to transfer bees from the common hives, during all the season that the weather is warm enough to permit them to fly; and yet to be able to guarantee that they will receive no serious damage by the change.

On the 10th of November, 1852, in the latitude of Northern Massachusetts, I transferred a colony which wintered in good health, and which now, May, 1853, promises to make an excellent stock. The day was warm, but after the operation was completed, the weather suddenly became cold, and as the bees were not able to leave the hive in order to obtain the water necessary for repairing their comb, they were supplied with that indispensable article. They went to work *very* busily, and in a short time mended up their combs and attached them firmly to the frames.

The transfer may be made of any healthy colony, and if they are strong in numbers, and the hive is well provisioned, and the weather is not too cool when the operation is attempted, they will scarcely feel the change. If the weather should be too chilly, it will be found almost impossible to make a colony leave its old hive, and if the combs are cut out, and the bees removed upon them, large numbers of them will take wing, and becoming chilled, will be unable to join their companions, and so will perish.

The process of transferring bees to my hives, is performed as follows. Let the old hive be shut up and well drummed[25] and the bees, if possible, be driven into an upper box. If they will not leave the hive of their own accord, they will fill themselves, and when it is ascertained that they are determined, if they can help it, not to be tenants at will, the upper box must be removed, and the bees gently sprinkled, so that they may all be sure to have nothing done to them on an empty stomach. If possible, an end of the old box parallel with the combs, must be pried off, so that they may be easily cut out. An old hive or box should stand upon a sheet, in place of the removed stock, and as fast as a comb is cut out, the bees should be shaken from it, upon the sheet; a wing or anything soft, will often be of service in brushing off the bees. Remember that they must not be hurt. If the weather is so pleasant that many bees from other hives, are on the wing, great care must be taken to prevent them from robbing. As fast therefore as the bees are shaken from the combs, these should be put into an empty hive or box, and covered with a cloth, or set in some place where they will not be disturbed. As soon as all the combs have been removed, the Apiarian should proceed to select and arrange them for his new hive. If the transfer is made late in the season, care must be taken, of course, to give the bees combs containing a generous allowance of honey for their winter supplies; together with such combs as have brood, or are best fitted for the rearing of workers. All coarse combs except such as contain the honey which they need, should be rejected. Lay a frame upon a piece of comb, and mark it so as to be able to cut it a trifle larger, so that it will just *crowd* into the frame, to remain in its place until the bees have time to attach it. If the size of the combs is such, that some of them cannot be cut so as to fit, then cut them to the best advantage, and after putting them into the frames, wind some thread around the upper and lower slats of the frame, so as to hold the combs in their place, until the bees can fasten them. If however, any of the combs which do not fit, have no honey in them, they may be fastened very easily, by dipping their upper edges into melted rosin. When the requisite number of combs are put into the frames, they should be placed in the new hive, and slightly fastened on the rabbets with a mere touch of paste, so as to hold them firmly in their places; this will be the more necessary if the transfer is made so late in the season that the bees cannot obtain the propolis necessary to fasten them, themselves.

As soon as the hive is thus prepared, let the temporary box into which the bees have been driven, be removed, and their new home put in its place. Shake out now the bees from the box, upon a sheet in front of this hive, and the work is done; bees, brood, honey, bee-bread, empty combs and all, have been nicely moved, and without any more serious loss than is often incurred by any other moving family, which has to mourn over some broken crockery, or other damage done in the necessary work of establishing themselves in a new home! If this operation is performed at a season of the year when there is much brood in the hive, and when the weather is cool, care must be taken not to expose the brood, so that it may become fatally chilled.

The best time for performing it, is late in the Fall, when there is but little brood in the hive; or about ten days after the voluntary or forced departure of a first swarm from the old stock. By this time, the brood left by the old queen, will all be sealed over, and old enough to bear exposure, especially as the weather, at swarming time, is usually quite warm. A temperature, not lower than 70°, will do them no harm, for if exposed to such a temperature, they will hatch, even if taken from the bees.

I have spoken of the *best* time for performing this operation. It may be done at any season of the year, when the bees can fly without any danger of being chilled, and I should not be afraid to attempt it, in mid-winter, if the weather was as warm as it sometimes is. Let me here earnestly caution all who keep bees, against meddling with them when the weather is cool. Irreparable mischief is often done to them at such times; they are tempted to fly, and thus perish from the cold, and frequently they become so much excited, that they cannot retain their fæces, but void them among the combs. If nothing worse ensues, they are disturbed when they ought to be in almost death-like repose, and are thus tempted to eat a much larger quantity of food than they would otherwise have needed. Let the Apiarian remember that not a single unnecessary motion should be required of a single bee: for all this, to say nothing else, involves a foolish waste of food.

In all operations involving the transferring of bees, it is exceedingly desirable that the new hives to which they are transferred should be put, as near as possible, where the old ones stood. If other colonies are in close proximity, the bees may be tempted to enter the wrong hives, if their position is changed only a little; they are almost sure to do this if the others resemble more closely than the new one, their former habitation. If will be often advisable, to transport to the distance of one or two miles, the stocks which are to be transferred; so that the operation may be performed to the best advantage. In a few weeks they may be brought back to the Apiary. In hiving swarms, and transferring stocks, care must be taken to prevent the bees from getting mixed with those of other colonies. If this precaution is neglected many bees will be lost by joining other stocks, where they may be kindly welcomed, or may at once be put to death. It is exceedingly difficult, to tell before hand, what kind of a reception strange bees will meet with, from a colony which they attempt to join. In the working season they are much more likely to be well received, than at any other time, especially if they come loaded with honey: still new swarms full of honey, that attempt to enter other hives, are often killed at once. If a colony which has an unimpregnated queen seeks to unite with another which has a fertile one, then almost as a matter of course they are destroyed! If by moving their hive, or in any other way, bees are made to enter a hive containing an unimpregnated queen, they will often destroy her, if they came from a family which was in possession of a fertile one! If any thing of this kind is ever attempted, the queen ought first to be confined in a queen cage. If while attempting a transfer of the bees to a new hive, I am apprehensive of robbers attacking the combs, or am pressed for want of time, I put only such combs as contain brood into the frames, and set the others in a safe place. The bees are now at once allowed to enter their new hive, and the other combs are given to them at a more convenient time. The whole process of transferal need not occupy more than an hour, and in some cases it can be done in fifteen minutes. If the weather is hot, the combs must not be exposed at all to the heat of the sun.

Until I had tested the feasibility of transferring bees from the old hives, by means of my frames, I felt strongly opposed to any attempt to dislodge them from their previous habitation. If they are transferred in the usual way, it must be done when the combs are filled with brood; for if delayed until late in the season, they will have no time to lay in a store of provision against the Winter. Who can look without disgust, upon the wanton destruction of thousands of their young, and the silly waste of comb, which can be replaced only by the consumption of large quantities of honey? In the great majority of such cases, the transfer, unless made about the swarming season, and *previous* to the issue of the first swarm, will be an entire failure, and if made before, at best only one colony is obtained, instead of the two, which are secured on my plan. I never advise the transfer of a colony into *any* hive, unless their combs can be transferred with them, nor do I advise any except practical Apiarians, to attempt to transfer them even to my hives. But what if a colony is so old that its combs can only breed dwarfs? When I find such a colony, I shall think it worth while to give specific directions as to how it should be managed. The truth is, that of all the many mistakes and impositions which have disgusted multitudes with the very sound of "patent hive," none has been more fatal than the notion that an old colony of bees could not be expected to prosper. Thousands of the very best stocks have been wantonly sacrificed to this Chimera; and so long as bee-keepers instead of studying the habits of the bee, prefer to listen to the interested statements of ignorant, or enthusiastic, or fraudulent persons, thousands more will suffer the same fate. As to old stocks, the prejudice against them is just as foolish as the silly notions of some who imagine that a woman is growing old, long before she has reached her prime. Many a man of mature years who has married a girl or a child, instead of a woman, has often had both time enough, and cause enough to lament his folly.

It cannot be too strongly urged upon all who keep bees, either for love or for money, to be exceedingly cautious in trying any new hive, or new system of management. If you are ever so well satisfied that it will answer all your expectations, enter upon it, at first, only on a small scale; then, if it fulfills all its promises, or if *you* can make it do so, you may safely adopt it: at all events, you will not have to mourn over large sums of money spent for nothing, and numerous powerful colonies entirely destroyed. "Let well enough alone," should, to a great extent, be the motto of every prudent bee-keeper. There is, however, a golden mean between that obstinate and stupid conservatism which tries nothing new, and, of course, learns nothing new, and that craving after mere novelty, and that rash experimenting on an extravagant scale, which is so characteristic of a large portion of our American people. It would be difficult to find a better maxim than that which is ascribed to David Crockett; "*Be sure you're right, then go ahead.*"

What old bee-keeper has not had abundant proof that stocks eight or ten years old, or even older, are often among the very best, in his whole Apiary, always healthy and swarming with almost unfailing regularity! I have seen such hives, which for more than fifteen years, have scarcely failed, a single season, to throw a powerful swarm. I have one now ten years old, in admirable condition, which a few years ago, swarmed three times, and the first swarm sent off a colony the same season. All these swarms were so early that they gathered ample supplies of honey, and wintered without any assistance!

I have already spoken of old stocks flourishing for a long term of years in hives of the roughest possible construction; and I shall now in addition to my previous remarks assign a new reason for such unusual prosperity. Without a single exception, I have found one or both of two things to be true, of every such hive. Either it was a very large hive, or else if not of unusual size, it contained a large quantity of worker-comb. No hive which does not contain a good allowance of regular comb of a size adapted to the rearing of workers, can ever in the nature of things, prove a valuable stock hive. Many hives are so full of drone combs that they breed a cloud of useless consumers, instead of the thousands of industrious bees which ought to have occupied their places in the combs. It frequently happens that when bees are put into a new hive, the honey-harvest is at its height, and the bees finding it difficult to build worker comb fast enough to hold their gatherings, are tempted to construct long ranges of drone comb to receive their stores. In this way, a hive often contains so small an allowance of worker-comb, that it can never flourish, as the bees refuse to pull down, and build over any of their old combs. All this can be easily remedied by the use of the movable comb hive.

PROCURING BEES TO START AN APIARY.

A person ignorant of bees, must depend in a very great measure, on the honesty of those from whom he purchases them. Many stocks are not worth accepting as a gift: like a horse or cow, incurably diseased, they will only prove a bill of vexatious expense. If an inexperienced person wishes to commence bee-keeping, I advise him, by all means, to purchase a new swarm of bees. It ought to be a large and early one. Second swarms and all late and small first swarms, ought never to be purchased by one who has no experience in Apiarian pursuits. They are very apt, in such hands, to prove a failure. If all bee-keepers were of that exemplary class of whom the Country Curate speaks, it would be perfectly safe to order a swarm of any one keeping a stock of bees. This however, is so far from being true, that some offer for sale, old stocks which are worthless, or impose on the ignorant, small first swarms, and second and even third swarms, as prime swarms worth the very highest market price. If the novice purchases an old stock, he will have the perplexities of swarming, &c., the first season, and before he has obtained any experience. As it may, however, be sometimes advisable that this should be done, unless he makes his purchase of a man known to be honest, he should select his stock himself, at a period of the day when the bees, in early Spring, are busily engaged in plying their labors. He should purchase a colony which is very actively engaged in carrying in bee-bread, and which, from the large number going in and out, undoubtedly contains a vigorous population. The hive should be removed at an hour when the bees are all at home. It may be gently inverted, and a coarse towel placed over it, and then tacked fast, when the bees are shut in. Have a steady horse, and before you start, be very sure that it is *impossible* for any bees to get out. Place the hive on some straw, in a wagon that has easy springs, and the bees will have plenty of air, and the combs, from the inverted position of the hive, will not be so liable to be jarred loose. Never purchase a hive which contains much comb just built; for it will be next to impossible to move it, in warm weather, without loosening the new combs. If a new swarm is purchased, it may be brought home as follows. Furnish the person on whose premises it is to be hived, with a box holding at the very least, a cubic foot of clear contents. Let the bottom-board of this temporary hive be clamped on both ends, the clamps being about two inches wider than the thickness of the board, so that when the hive is set on the bottom-board, it will slip in between the upper projections of the clamps, and be kept an inch from the ground, by the lower ones, so that air may pass under it. There should be a hole in the bottom-board, about four inches in diameter, and two of the same size in the opposite sides of the box, covered with wire gauze, so that the bees may have an abundance of air, when they are shut up. Three parallel strips, an inch and a half wide, should be nailed, about one third of the way from the top of the temporary hive, at equal distances apart, so that the bees may have every opportunity to cluster; a few pieces of old comb, fastened strongly in the top with melted rosin, will make the bees like it all the better. A handle made of a strip of leather, should be nailed on the top. Let the bees be hived in this box, and kept well shaded; at evening, or very early next morning, the temporary hive which was propped up, when the bees were put into it, may be shut close to its bottom-board, and a few screws put into the upper projection of the clamps, so as to run through into the ends of the box. In such a box, bees may be safely transported, almost any reasonable distance: care being taken not to handle them roughly, and never to keep them in the sun, or in any place where they have not sufficient air. If the box is too small, or sufficient ventilators are not put in, or if the bees are exposed to too much heat, they will be sure to suffocate. If the swarm is unusually large, and the weather excessively warm, they ought to be moved at night. Unless great care is taken in moving bees, in very hot weather, they will be almost sure to perish; therefore always be *certain* that they have an abundance of air. If they appear to be suffering for want of it, especially if they begin to fall down from the cluster, and to lie in heaps on the bottom-board, they should immediately be carried into a field or any convenient place, and at once be allowed to fly: in such a case they cannot be safely moved again, until towards night. This will never be necessary if the box is large enough, and suitably ventilated.

I have frequently made a box for transporting new swarms, out of an old tea-chest. When a new swarm is brought in this way to its intended home, the bottom-board may be unscrewed, and the bees transferred at once, to the new hive; In some cases, it may be advisable to send away the new hive. In this case, if one of my hives is used, the spare honey-board should be screwed down, and all the holes carefully stopped, except two or three which ought to have some ventilators tacked over them: the frames should be fastened with a little paste, so that they will not start from their place, and after the bees are hived, the blocks which close the entrance should be screwed down to their place, keeping them however, a trifle less than an eighth of an inch from the entrance, so as to give the bees all the air which they need. I very much prefer sending a box for the bees: one person can easily carry two such boxes, each with a swarm of bees; and if he chooses to fasten them to two poles, or to a very large hoop, he may carry four, or even more.

If the Apiarian wishes, to be sure the first season, of getting some honey from his bees, he will do well to procure two good swarms, and put them both into one hive. To those who do not object to the extra expense, I strongly recommend this course. Not infrequently, they will in a good season, obtain in spare honey from their doubled swarm, an ample equivalent for its increased cost: at all events, such a powerful swarm lays the foundations of a flourishing stock, which seldom fails to answer all the reasonable expectations of its owner. If the Apiary is commenced with swarms of the current season, and they have an abundance of spare room in the upper boxes, there will be no swarming, that season, and the beginner will have ample time to make himself familiar with his bees, before being called to hive new swarms, or to multiply colonies by artificial means.

Let no inexperienced person commence bee-keeping on a large scale; very few who do so, find it to their advantage, and the most of them not only meet with heavy losses, but abandon the pursuit in disgust. By the use of my hives, the bee-keeper can easily multiply very rapidly, the number of his colonies, as soon as he finds, not merely that money can be made by keeping bees, but *that he can make it*. While I am certain that more money can be made by a careful and experienced bee-keeper in a good situation, from a given sum invested in an Apiary, than from the same money invested in any other branch of rural economy, I am equally certain that there is none in which a careless or inexperienced person would be more sure to find his outlay result in an almost entire loss. An Apiary neglected or mismanaged, is far worse than a farm overgrown with weeds, or exhausted by ignorant tillage: for the land is still there, and may, by prudent management, soon be made again to blossom like the rose; but the bees, when once destroyed, can never be brought back to life, unless the poetic fables of the Mantuan Bard, can be accepted as the legitimate results of actual experience, and swarms of bees, instead of clouds of filthy flies, can once more be obtained from the carcases of decaying animals! I have seen an old medical work in which Virgil's method of obtaining colonies of bees from the putrid body of a cow slain for this special purpose, is not only credited, but minutely described.

A large book would hardly suffice to set forth all the superstitions connected with bees. I will refer to one which is very common and which has often made a deep impression upon many minds. When any member of a family dies, the bees are believed to be aware of what has happened, and the hives are by some dressed in mourning, to pacify their sorrowing occupants! Some persons imagine that if this is not done, the bees will never afterwards prosper, while others assert, that the bees often take their loss so much to heart, as to alight upon the coffin whenever it is exposed! An intelligent clergyman on reading the sheets of this work, stated to me that he had always refused to credit this latter fact, until present at a funeral where the bees gathered in such large numbers upon the coffin, as soon as it was brought out from the house, as to excite considerable alarm. Some years after this occurrence, being engaged in varnishing a table, and finding that the bees came and lit upon it, he was convinced that the love of varnish, instead of sorrow or respect for the dead, was the occasion of their gathering round the coffin! How many superstitions in which often intelligent persons most firmly confide, might if all the facts were known, be as easily explained.

Before closing this Chapter, I must again strongly caution all inexperienced bee-keepers, against attempting to transfer colonies from an old hive. I am determined that if any find that they have made a wanton sacrifice of their bees, they shall not impute their loss to my directions. If they persist in making the attempt, let them, by all means, either do it at break of day, before the bees of other hives will be induced to commence robbing; or better still, let them do it not only early in the morning, but let them carry the hive on which they intend to operate, to a very considerable distance from the vicinity of the other hives, and entirely out of sight of the Apiary. I prefer myself this last plan, as I then run no risk of attracting other bees to steal the honey, and acquire mischievous habits.

The bee-keeper is very often reminded by the actions of his bees of some of the worst traits in poor human nature. When a man begins to sink under misfortunes, how many are ready not simply to abandon him, but to pounce upon him like greedy harpies, dragging, if they can, the very bed from under his wife and helpless children, and appropriating all which by any kind of maneuvering, they can possibly transfer to their already overgrown coffers! With much the same spirit, more pardonable to be sure in an insect, the bees from other hives, will gather round the one which is being broken up, and while the disconsolate owners are lamenting over their ruined prospects, will, with all imaginable rapacity and glee, bear off every drop which they can possibly seize.

CHAPTER XV. ROBBING.

Bees are exceedingly prone to rob each other, and unless suitable precautions are used to prevent it, the Apiarian will often have cause to mourn over the ruin of some of his most promising stocks. The moment a departure is made from the old-fashioned mode of managing bees, the liability to such misfortunes is increased, unless all operations are performed by careful and well informed persons.

Before describing the precautions which I successfully employ, to guard my colonies from robbing each other, or from being robbed by bees from a strange Apiary, I shall first explain under what circumstances they are ordinarily disposed to plunder each other. Idleness is with bees, as well as with men, a most fruitful mother of mischief. Hence, it is almost always when they are doing nothing in the fields, that they are tempted to increase their stores by dishonest courses. Bees are, however, much more excusable than the lazy rogues of the human family; for the *bees* are idle, not because they are indisposed to work, but because they can find nothing to do. Unless there is some gross mismanagement, on the part of their owner, they seldom attempt to live upon stolen sweets, when they have ample opportunity to reap the abundant harvests of honest industry. In this chapter, I shall be obliged, however much against my will, to acknowledge that some branches of morals in my little friends, need very close watching, and that they too often make the lowest sort of distinction, between "mine and thine." Still I feel bound to show that when thus overcome by temptation, it is almost always, under circumstances in which their careless owner is by far the most to blame.

In the Spring, as soon as the bees are able to fly abroad, "innatus urget amor habendi," as Virgil has expressed it; that is, they begin to feel the force of an innate love of honey-getting. They can find nothing in the fields, and they begin at once, to see if they cannot appropriate the spoils of some weaker hive. They are often impelled to this, by the pressure of immediate want, or the salutary dread of approaching famine: but truth obliges me to confess that not infrequently some of the strongest stocks, which have more than they would be able to consume, even if they gathered nothing more for a whole year, are the most anxious to prey upon the meager possessions of some feeble colony. Just like some rich men who have more money than they can ever use, urged on by the insatiable love of gain, "oppress the hireling in his wages, the widow and the fatherless," and spin on all sides, their crafty webs to entrap their poorer neighbors, who seldom escape from their toils, until every dollar has been extracted from them, and as far as their worldly goods are concerned, they resemble the skins and skeletons which line the nest of some voracious old spider.

When I have seen some powerful hive of the kind just described, condemned by its owner, in the Fall, to the sulphur pit, or deprived unexpectedly of its queen, its stores plundered, and its combs eaten up by the worms, I have often thought of the threatenings which God has denounced against those who make dishonest gains "their hope, and say unto the fine gold, Thou art my confidence."

In order to prevent colonies from attempting to rob, I always examine them in the Spring, to ascertain that they have honey and are in possession of a fertile queen. If they need food they are supplied with it, (see Chapter on Feeding,) and if they are feeble or queenless, they are managed according to the directions previously given. Bees seem to have an instinctive perception of the weakness of a colony, and like the bee-moth, they are almost certain to attack such stocks, especially when they have no queen. Hence I can almost always tell that a colony is queenless, by seeing robbers constantly attempting to force an entrance into it.

It requires some knowledge of the habits of bees, to tell from their motions, whether they are flying about a strange hive with some evil intent, or whether they belong to the hive before which they are hovering. A little experience however, will soon enable us to discriminate between the honest inhabitants of a hive, and the robbers which so often mingle themselves among the crowd. There is an unmistakable air of roguery about a thieving bee, which to the observing Apiarian, proclaims the nature of his calling, just as truly as the appearance of a pickpocket in a crowd, enables the experienced police officer to distinguish him from the honest folks, on whom he intends to exercise his skill.

There is a certain sneaking look about a rogue of a bee, almost indescribable, and yet perfectly obvious. It does not alight on the hive, and boldly enter at once like an honest bee which is carrying home its load. If they could only assume such an appearance of transparent honesty, they would often be allowed by the unsuspecting door-keepers to enter unquestioned, to see all the sights within, and to help themselves to the very fat of the land. But there is a sort of nervous haste, and guilty agitation in all their movements: they never alight boldly upon the entrance board, or face the guards which watch the passage to the hive; they know too well that if caught and overhauled by these trusty guardians of the hive, their lives would hardly be worth insuring; hence their anxiety to glide in, without touching one of the sentinels. If detected, as they have no password to give, (having a strange smell,) they are very speedily dealt with, according to their just deserts. If they can only effect a secret entrance, those within take it for granted that all is right, and seldom subject them to a close examination.

Sometimes bees which have lost their way, are mistaken by the inexperienced, for robbers; there is however, a most marked distinction between the conduct of the two. The arrant rogue when caught, attempts with might and main, to pull away from his executioners, while the poor bewildered unfortunate shrinks into the smallest compass, like a cowed dog, and submits to whatever fate his captors may see fit to award him.

The class of dishonest bees which I have been describing, may be termed the "Jerry Sneaks" of their profession, and after they have followed it for some time, they lose all disposition for honest pursuits, and assume a hang-dog sort of look, which is very peculiar. Constantly employed in creeping into small holes, and daubing themselves with honey, they often lose all the bright feathers and silky plumes which once so beautifully adorned their bodies, and assume a smooth and almost black appearance; just as the hat of the thievish loafer, acquires a "seedy" aspect, and his garments, a shining and threadbare look. Dzierzon is of opinion that the black bees which Huber describes, as being so bitterly persecuted by the rest, are nothing more than these thieving bees. I call them old convicts, dressed in prison garments, and incurably given up to dishonest pursuits.

Bees sometimes act the part of highway robbers; some half dozen or more of them, will waylay and attack a poor humble-bee which is returning with a sack full of honey to his nest, like an honest trader, jogging home with a well filled purse. They seize the poor bee, and give him at once to understand that they must have the earnings of his industry. They do not slay him. Oh no! they are much too selfish to endanger their own precious persons; and even if they could kill him, without losing their weapons, they would still be unable to extract his sweets from the deep recesses of his honey bag: they therefore begin to bite and teaze him, after the most approved fashion, all the time singing in his ears, "not your money," but, "your honey or your life;" until utterly discouraged, he delivers up his purse, by disgorging his honey from its capacious receptacle. The graceless creatures cry "hands off," and release him at once, while they lick up his spoils and carry it off to their home.

The remark is frequently made that were rogues to spend half as much time and ingenuity in gaining an honest living, as they do, in seeking to impose upon their fellow-men, their efforts would often be crowned with abundant success. Just so of many a dishonest bee. If it only knew its true interests, it would be safely roving the smiling fields, in search of honey, instead of longing for a tempting and yet dangerous taste of forbidden sweets.

Bees sometimes carry on their depredations on a more magnificent scale. Having ascertained the weakness of some neighboring colony, through the sly intrusions of those who have entered the hive to spy out all "the nakedness of the land," they prepare themselves for war, in the shape of a pitched battle. The well-armed warriors sally out by thousands, to attack the feeble hive against which they have so unjustly declared a remorseless warfare. A furious onset is at once made, and the ground in front of the assaulted hive is soon covered with the dead and dying bodies of innumerable victims. Sometimes the baffled invaders are compelled to sound a retreat; too often however, as in human contests, right proves but a feeble barrier against superior might; the citadel is stormed, and the work of rapine and pillage forthwith begins. And yet after all, matters are not nearly so bad, as at first they seem to be. The conquered bees, perceiving that there is no hope for them in maintaining the unequal struggle, submit themselves to the pleasure of the victors; nay more, they aid them in carrying off their own stores, and are immediately incorporated into the triumphant nation! The poor mother however, is left behind in her deserted home, some few of her children which are faithful to the last, remaining with her, to perish by her side, amid the sad ruins of their once happy home!

If the bee-keeper is unwilling to have his bees so demoralized, that their value will be seriously diminished, he will be exceedingly careful to do all that he possibly can to prevent them from robbing each other. He will see that all queenless colonies are seasonably broken up in the Spring, and all weak ones strengthened, and confined to a space which they can warm and defend. If once his bees get a taste of forbidden sweets, they will seldom stop until they have tested the strength of every stock, and destroyed all that they possibly can. Even if the colonies are able to defend themselves, many bees will be lost in these encounters, and a large waste of time will invariably follow; for bees whether engaged in attempting to rob, or in battling against the robbery of others, are, to a very great extent, cut off both from the disposition and the ability to engage in useful labors. They are like nations that are impoverished by mutual assaults on each other: or in which the apprehension of war, exerts a most blighting influence upon every branch of peaceful industry.

I place very great reliance on the movable blocks which guard the entrance to my hive, to assist colonies in defending themselves against robbing bees, as well as the prowling bee-moth. These blocks are triangular in shape, and enable the Apiarian to enlarge or contract the entrance to the hive, at pleasure. In the Spring, the entrance is kept open only about two inches, and if the colony is feeble, not more than half an inch. If there is any sign of robbers being about, the small colonies have their entrances closed, so that only a single bee can go in and out at once. As the bottom-board slants forwards, the entrance is on an inclined plane, and the bees which defend it, have a very great advantage over those which attack them; the same in short, that the inhabitants of a besieged fortress would have in defending a pass-way similarly constructed. As only one bee can enter at a time, he is sure to be overhauled, if he attempts ever so slyly to slip in: his credentials are roughly demanded, and as he can produce none, he is at once delivered over to the executioners. If an attempt is made to gain admission by force, then as soon as a bee gets in, he finds hundreds, if not thousands, standing in battle array, and he meets with a reception altogether too warm for his comfort. I have sometimes stopped robbing, even after it had proceeded so far that the assaulted bees had ceased to offer any successful resistance, by putting my blocks before the entrance, and permitting only a single bee to enter at once: the dispirited colony have at once recovered heart, and have battled so stoutly and successfully, as to beat off their assailants.

When bees are engaged in robbing a hive, they will often continue their depredations to as late an hour as possible, and not infrequently some of them return home so late with their ill-gotten spoils, that they cannot find the entrance to their own hive. Like the wicked man who "deviseth mischief on his bed, and setteth himself in a way that is not good," they are all night long, meditating new violence, and with the very first peep of light, they sally out to complete their unlawful doings.

Sometimes the Apiarian may be in doubt whether a colony is

being robbed or not, and may mistake the busy numbers that arrive and depart, for the honest laborers of the hive; but let him look into the matter a little more closely, and he will soon ascertain the true state of the case: the bees that enter, instead of being heavily laden, with bodies hanging down, unwieldy in their flight, and slow in all their movements, are almost as hungry looking as Pharaoh's lean kine, while those that come out, show by their burly looks, that like aldermen who have dined at the expense of the City, they are filled to their utmost capacity.

If the Apiarian wishes to guard his bees against the fatal propensity to plunder each other, he must be exceedingly careful not to have any combs filled with honey unnecessarily exposed. An ignorant or careless person attempting to multiply colonies on my plan, will be almost sure to tempt his bees to rob each other. If he leaves any of the combs which he removes, so that strange bees find them, they will, after once getting a taste of the honey, fly to any hive upon which he begins to operate, and attempt to appropriate a part of its contents. I have already stated that when they can find an abundance of food in the fields, bees are seldom inclined to rob; for this reason, with proper precautions, it is not difficult to perform all the operations which are necessary on my plan of management, at the proper season, without any danger of demoralizing the bees. If however, they are attempted when honey cannot be obtained, they should be performed with extreme caution, and early in the morning, or late in the evening; or if possible, on a day when the bees are not flying out from their hives. I have sometimes seen the most powerful colonies in an Apiary, either robbed and destroyed, or very greatly reduced in numbers, by the gross carelessness or ignorance of their owner. He neglects to examine his hives at the proper season, and the bees begin to rob a weak or queenless stock: as soon as they are at the very height of their nefarious operations, he attempts to interfere with their proceedings, either by shutting up the hive, or by moving it to a new place. The air is now filled with greedy and disappointed bees, and rather than fail in obtaining the expected treasures, they assail with almost frantic desperation, some of the neighboring stocks: in this way, the most powerful colonies are sometimes utterly ruined, or if they escape, thousands of bees are slain in defending their treasures, and thousands more of the assailants meet with the same untimely end.

If the Apiarian perceives that one of his colonies is being robbed, he should at once contract the entrance, so that only a single bee can get in at a time; and if the robbers still persist in entering, he must close it entirely. In a few minutes the outside of the hive will be black with the greedy cormorants, and they will not abandon it, until they have explored every crevice, and attempted to force themselves through even the smallest openings. Before they assail a neighboring colony, they should be sprinkled with cold water, and then instead of feeling courage for new crimes, they will be glad to escape, thoroughly drenched, to their proper homes. Unless the bees that are shut up can, as in my hives, have an abundance of air, it will be necessary to carry them at once into a dark and cool place. Early next morning the condition of the hive should be examined, and the proper remedies if it is weak or queenless should be applied; or if its condition is past remedy, it should at once be broken up, and the bees united to another stock.

I have been credibly informed of an exceedingly curious kind of robbing among bees. Two colonies, both in good condition, seemed determined to appropriate each other's labors: neither made any resistance to the entrance of the plundering bees; but each seemed too busily intent upon its own dishonest gains, to notice[26] that the work of subtraction kept pace with that of addition. An intelligent Apiarian stated to me this singular fact as occurring in his own Apiary. This is a very near approximation to the story of the Kilkenny cats. Alas! that there should be so much of equally short-sighted policy among human beings; individuals, communities and nations seeking often to thrive by attempting to prey upon the labors of others, instead of doing all that they can, by industry and enterprise, to add to the common stock. I have never, in my own experience, met with an instance of such silly pilfering as the one described; but I have occasionally known bees to be carrying on their labors, while others were stealing more than the occupants of the hive were gathering, without their being aware of it.

CHAPTER XVI. DIRECTIONS FOR FEEDING BEES.

Few things in the practical department of the Apiary, are more impor- tant and yet more shamefully neglected, or grossly mismanaged, than the feeding of bees. In order to make this subject as clear as possible, I shall begin with the Spring examination of the hives, and furnish suitable directions for feeding during the whole season in which it ought to be attempted. In the movable comb hives, the exact condition of the bees with regard to stores, may be easily ascertained as soon as the weather is warm enough to lift out the frames. In the common hives, this can sometimes be ascertained from the glass sides; but often no reliable information can be obtained. Even if the weight of the hive is known, this will be no sure criterion of the quantity of honey it contains. The comb in old hives, is often very thick, and of course, unusually heavy; while vast stores of use- less bee-bread have frequently been accumulated, which entirely deceive the Apiarian, who attempts to judge of the resources of a hive from its weight alone. On my system of bee-culture, such an injurious surplus of bee-bread, is easily prevented;

If the bee-keeper ascertains or even suspects, in the Spring, that his bees have not sufficient food, he must at once supply them with what they need. Bees, at this season of the year, consume a very large quantity of honey: they are stimulated to great activity by the returning warmth, and are therefore compelled to eat much more than when they were almost dormant among their combs. In addition to this extra demand, they are now engaged in rearing thousands of young, and all these require a liberal supply of food. Owing to the inexcusable neglect of many bee-keepers, thousands of swarms perish annually after the Spring has opened, and when they might have been saved, with but little trouble or expense. Such abominable neglect is incomparably more cruel than the old method of taking up the bees with sulphur; and those who are guilty of it, are either too ignorant or too careless, to have any thing to do with the management of bees. What would be thought of a farmer's skill in his business, who should neglect to provide for the wants of his cattle, and allow them to drop down lifeless in their stalls, or in his barn-yard, when the fields, in a few weeks, will be clothed again with the green mantle of delightful Spring! If any farmer should do this, when food might easily be purchased, and should then, while engaged in the work of skinning the skeleton carcasses of his neglected herd, pretend that he could not afford to furnish, for a few weeks, the food which would have kept them alive, he would not be a whit more stupid than the bee-keeper attempting to justify himself on the score of economy, while engaged in melting down the combs of a hive, starved to death, after the Spring has fairly opened! Let such a person blush at the pretence that he could not afford to feed his bees, the few pounds of sugar or honey, which would have saved their lives, and enabled them to repay him tenfold for his prudent care.

I always feed my bees a little, even if I know that they have enough and to spare. There seems to be an intimate connection between the getting of honey, and the rapid increase of breeding, in a hive; and the taste of something sweet, however small, to be added to their hoards, exerts a very stimulating effect upon the bees; a few spoonsfull a day, will be gratefully received, and will be worth much more to a stock of bees in the Spring, than at any other time.

By judicious early feeding, a whole Apiary may be not only encouraged to breed much faster than they otherwise would have done; but they will be inspired with unusual vigor and enterprise, and will afterwards increase their stores with unusual rapidity. Great caution must be exercised in supplying bees at this time with food, both to prevent them from being tempted to rob each other, or to fill up with honey, the cells which ought to be supplied with brood. Only a small allowance should be given to them, and this from time to time, unless they are destitute of supplies; and as soon as they begin to gather from the fields, the feeding should be discontinued. Feeding, intended merely to encourage the bees, and to promote early breeding, may be done in the open air. No greater mistake can be made than to feed largely at this season of the year. The bees take, to be sure, all that they can, and stow it up in their cells, but what is the consequence? The honey which has been fed to them, fills up their brood combs, and the increase of population is most seriously interfered with; so that often when stocks which have not been over-fed, are prepared not only to fill all the store combs in their main hive, but to take speedy possession of the spare honey boxes, a colony imprudently fed, is too small in numbers, to gather even as much as the one which was not fed at all! The inexperienced Apiarian has thus often made a worse use of his honey than he would have done, if he had actually thrown it away! while all the time, he is deluding himself with the vain expectation of reaping some wonderful profits, from what he considers an improved mode of managing bees.

Such conduct in its results, appears to me very much like the noxious influences under which too many of the children of the rich are so fatally reared. With every want gratified, pampered and fed to the very full, how often do we see them disappoint all the fond expectations of parents and friends, their money proving only a curse, while not infrequently beggared in purse, and bankrupt in character, they prematurely sink to an ignoble or dishonored grave. Think of it, ye who are slaving in the service of Mammon, that ye may leave to your sons, the overgrown wealth which usually proves a legacy of withering curses, while you neglect to train them up in those habits of stern morality and steady industry, and noble self-reliance, without which the wealth of Crœsus would be but a despicable portion! Think of it, as you contrast its results in the bitter experience of thousands, with the happier influences under which so many of our noblest men in Church and State, have been nurtured and developed, and then pursue your sordid policy, if you can. "There is that withholdeth" from good objects, "more than is meet, and it tendeth to poverty:" yes, to poverty of Christian virtue and manliness, and of those "treasures" which we are all entreated by God himself, to "lay up" in the store-house of Heaven. Call your narrow-mindedness and gross deficiencies in Christian liberality, nothing more than a natural love of your children, and an earnest desire to provide for your own household. Little fear there may be that *you* will ever incur the charge of being "worse than an infidel" on this point; but lay not on this account, any flattering unction to your souls; look within, and see if the base idolatry of gold has not more to do with your whole course of thinking and acting, than any love of wife or children, relatives or friends!

Another *sermon*! does some one exclaim? Would then that it might be to some of my readers a sermon indeed; "a word fitly spoken," "like apples of gold in pictures of silver."

The prudent Apiarian will always regard the feeding of bees, except the little, given to them by way of encouragement, as an evil to be submitted to, only when absolutely necessary; and will very much prefer to obtain his supplies from what Shakspeare has so beautifully termed the "merry pillage" of the blooming fields, than from the more costly stores of the neighboring grocery. If not engaged in the rapid increase of stocks, he will seldom see a season so unfavorable as to be obliged to purchase any food for his bees, unless he chooses to buy a cheaper article, to replace the choice honey of which he has deprived them. Just as soon as the Apiarian begins to multiply his stocks with very great rapidity, he must calculate upon feeding great quantities of honey to his bees. Before he attempts this on a large scale, let me once more give him a friendly caution, and if possible, persuade him to try very rapid multiplication with only a few of his stocks. In this way, he may experiment to his heart's content, without running the risk of seriously injuring his whole Apiary, and he may not only gain the skill and experience which will enable him subsequently to conduct a rapid increase, on a large scale, but may learn whether he is so situated that he can profitably devote to it the time and money which it will inevitably require.

Before giving directions for feeding bees when a rapid increase of colonies is aimed at, I shall first show in what manner the bee-keeper may feed his weak swarms in the Spring. If they are in the common hives, a small quantity of liquid honey may, at once be poured among the combs in which the bees are clustered: this may be done by pouring it into the holes leading to the spare honey boxes, but a much better way is to invert the hives, and pour in about a tea-cup full at once. The Apiarian can then see just where to pour it; he need not fear that the bees will be hurt by it; any more than a child will be either hurt or displeased by the sweets which adhere to its hands and face, as it feasts upon a generous allowance of the best sugar candy! When the bees have taken up all that has been poured upon them, the hive may be replaced, and the operation repeated in a few days: the oftener it is done, the better it will suit them. If the weather is sufficiently warm to allow the bees to fly without being chilled, the food may be put in some old combs, or in a feeder, and set in a sunny place, a rod or more from their hives. If placed too near, the bees may be tempted to rob each other. With my hives, I can pour the honey into some empty comb, and then put the frame containing it, directly into the hive; or I can set the feeder or honey in the comb, in the hive near the frames which contain the bees. I have already stated, that unless a colony can be supplied with a sufficient number of bees, it cannot be aided by giving it food. If the bees are not numerous enough to take charge of the eggs which the queen can lay, or at least, of a large number of them, they can seldom, unless they have a tropical season before them, increase rapidly enough to be of any value. If they are numerous enough to raise a great many young bees, but too few to build new comb, they must be fed very moderately, or they will be sure to fill up their brood comb with honey, instead of devoting themselves to the rapid increase of their numbers. If the Apiarian has plenty of empty worker comb which he can give them, he ought to supply them quite sparingly with honey, even when they are considerably numerous, in order to have them breed as fast as possible; not so sparingly however, as to prevent them from storing up any honey in sealed cells; or they will not be encouraged to breed as fast as they otherwise would. If he has no spare comb, and the hive is populous enough to build new comb, it must be supplied moderately, and by all means, *regularly* with the means of doing this; the object being to have comb building and breeding go together, so as mutually to aid each other. If the feeding is not regular, so as to resemble the natural supplies when honey is obtained from the blossoms, the bees will not use the food given to them, in building new comb, but chiefly in filling up all the cells previously built. If honey can be obtained regularly, and in sufficient quantities from the blossoms, the small colonies or nuclei will need no feeding until the failure of the natural supplies.

In all these operations, the main object should be to make every thing bend to the most rapid production of *brood*; give me the bees, and I can easily show how they may be fed, so as to make strong and prosperous stocks; whereas if the bees are wanting, every thing else will be in vain: just as a land where there are many stout hands and courageous hearts, although comparatively barren, will in due time, be made to "bud and blossom as the rose," while a second Eden, if inhabited by a scanty and discouraged population, must speedily be overgrown with briars and thorns.

If strong stocks are deprived of a portion of their combs, so that they cannot from natural sources, at once begin to refill all vacancies, they too must be fed.

I have probably said enough to show the inexperienced that the rapid multiplication of colonies is not a very simple matter, and that they will do well not to attempt it on a large scale. By the time the honey harvest ordinarily closes, all the colonies in the Apiaries of all except the skillful, ought to be both strong in numbers and in stores; at least the *aggregate* resources of the colonies should be such that when an equal division is made among them, there will be enough for them all. This may ordinarily be effected, and yet the number of the colonies be tripled in one season; and in situations where buckwheat is extensively cultivated, a considerable quantity of surplus honey may even then be frequently obtained from the bees. Early in the month of September, or better still, by the middle of August, if the colonies are sufficiently strong in numbers, I advise that if feeding is necessary to winter the bees, it should be thoroughly attended to. If delayed later than this, in the latitude of our Northern States, the bees may not have sufficient time to seal over the honey fed to them, and will be almost sure to suffer from dysentery, during the ensuing Winter. Unsealed honey, almost always, in cool weather, attracts moisture, and sours in the combs, and if the bees are compelled to feed upon it, they are very liable to become diseased. This is the reason why bees when fed with liquid honey, late in the Fall, or during the Winter, are almost sure to suffer from disease. A very interesting fact confirming these views as to the danger resulting from the use of sour food, has come under my notice this Spring. A colony of bees were fed for some time with suitable food, and appeared to be in perfect health, flying in and out with great animation. Their owner, on one occasion, before leaving for the day, gave them some molasses which was so *sour*, that it could not be used in the family. On returning, at evening, he was informed that the bees had been dropping their filth over every thing in the vicinity of the hive. On examining them, next day, they were all found dead on the bottom-board and among the combs! The acid food had acted upon them as a violent cathartic, and had brought on a complaint of which they all died in less than 24 hours: the hive was found to contain an ample allowance of honey and bee-bread.

If the Apiarian, on examining the condition of his stocks, finds that some have more than they need, and others not enough, his most prudent course will be to make an equitable division of the honey, among his different stocks. This may seem to be a very Agrarian sort of procedure, and yet it will answer perfectly well in the management of bees. Those that were helped, will not spend the next season in idleness, relying upon the same sort of aid; nor will those that were relieved of their surplus stores, remember the deprivation, and limit the extent of their gatherings to a bare competency. With men, most unquestionably, such an annual division, unless they were perfect, would derange the whole course of affairs, and speedily impoverish any community in which it might be attempted. I always prefer to take away a considerable quantity of honey from my stocks, which have too generous a supply, and to replace it with empty combs suitable for the rearing of workers; as I find that when bees have too much honey in the Fall, they do not ordinarily breed as fast in the ensuing Spring, as they otherwise would. A portion of this honey should be carefully put away in the frames, and kept in a close box, safe against all intruders, and where it will not be exposed to frost; so that if any colonies in the Spring, are found to be in want of food, they may easily be supplied.

In the Spring examination, if any colonies have too much honey, a portion of it ought by all means to be taken away. Such a deprivation, if judiciously performed, will always stimulate them to increased activity. Every strong stock, as soon as it can gather enough honey to construct comb, ought to have one or two combs which contain no brood removed, and their places supplied with empty frames, in order that they may be induced to exert themselves to the utmost. An empty frame inserted between full ones, will be replenished with comb very speedily, and often the combs removed will be so much clear gain. If at any time there is a sudden supply of honey, and the bees are reluctant to enter the boxes, or it is not probable that the supply will continue long enough to enable them to fill them, the removal of some of the combs from the main hive so as to have empty ones filled, will often be highly advantageous.

If in the Fall of the year, the bee-keeper finds that some of his colonies need feeding, and if they are not populous enough to make good stock hives in the ensuing Spring, then instead of wasting time and money on them, he should at once, break them up; They will seldom pay for the labor bestowed on them, and the bees will be much more serviceable, if added to other stocks. The Apiarian cannot be too deeply impressed with the important truth, that his profits in bee-keeping will all come from his *strong* stocks, and that if he cannot manage so as to have such colonies early, he had better let bee-keeping alone.

If liquid honey is fed to bees, it should always, be given to them seasonably, so that they may seal it over before the approach of cold weather. West India honey has for many years, been used to very good advantage, as a bee-feed. It should never be used in its raw state, as it is often filled with impurities, and is very liable to sour or candy in the cells, but should be mixed with about two parts of good white sugar, to three of honey and one of water, and brought to the boiling point; as soon as it begins to boil, it should be set to cool, and all the impurities will rise to the top, and may be skimmed off. If it is found to be too thick, a little more water may be added to it; it ought however, never to be made thinner than the natural consistence of good honey. Such a mixture will cost for a small quantity, about seven cents a pound, and will probably be found the cheapest liquid food, which can be given to bees. Brown sugar may be used with the honey, but the food will not be so good.

If one of my hives is used, the bee-keeper may feed his bees at the proper season, without using any feeder at all, or rather he may use the *bottom-board* of the hive as a feeder. On this plan, the bees should be fed at evening; so as to run no risk of their robbing each other. The hive which is to be fed, should have the front edge of its bottom-board elevated on a block, so as to slant *backwards*, and the honey should be poured into a small tin gutter inserted at the entrance; one such will answer for a whole Apiary, and may be made by bending up the edges of any old piece of tin. As the frames in my hive are kept about half an inch above the bottom-board, which is water-tight, the honey runs under them, and is as safe as in a dish, while the bees stand on the bottom of the frames, and help themselves. The quantity poured in, should of course, depend upon the size and necessities of the colony; no more ought to be given at one time than the bees can take up during the night, and the entrance to the hive ought always to be kept very small during the process of feeding, to prevent robber bees from getting in; a good colony will easily take up a quart. It is desirable to get through the feeding as rapidly as possible, as the bees are excited during the whole process, and consume more than they otherwise would; to say nothing of the demand made upon the time of the Apiarian, by feeding in small quantities. If the bees cannot, in favorable weather, dispose of at least a pint at one time, the colony must be too small to make it worth while to feed them, if they are in hives by which they can be readily united to stronger stocks.

If the bees have not a good allowance of comb, it will not, as a general rule, pay to feed them. This will be obvious to any one who reflects that at least 20 pounds of honey are required to elaborate one pound of wax. I know that this estimate may to some, appear enormous; but it is given as the result of very accurate experiments, instituted on a large scale, to determine this very point. The Country Curate says, "Having driven the population of four stocks, on the 5th of August, and united them together, I fed them with about 50 pounds of a mixture of sugar, honey, salt and beer, for about five weeks. At that time, the box was only

16 pounds heavier than when the bees were put into it." He then makes an estimate that at least 25 pounds of the mixture were consumed in making about half a pound of wax!! No one who has ever tried it, will undertake to feed bees for profit, when they are destitute of both comb and honey.

If the weather is cool when bees are fed, it will generally be necessary to resort to top feeding. For this, my hive is admirably adapted: a feeder may be put over one of the holes in the honey-board directly over the mass of the bees, into which the heat of the hive naturally arises, and where the bees can get at their food without any risk of being chilled. This is *always* the best place for a feeder, as the smell of the food is not so likely to attract the notice of robbing bees.

I shall here describe the way in which a feeder can at small expense, be made to answer admirably every purpose. Take any wooden box which will hold, say, at least one quart; make it honey-tight, by pouring into the joints the melted mixture, and brush the whole interior with the mixture, so that the honey may not soak into the wood. Make a float of thin wood, filled with quarter inch holes, with clamps nailed on the lower sides to prevent warping, and to keep the float from settling to the bottom of the box, so as to stick fast: it should have ample play, so that it may settle, as fast as the bees consume the honey. Tacks on the clamps will always be sure to prevent sticking. Before you waste any time in making small holes, for fear the bees will be drowned in the large ones, try a float made as directed. In one corner of the box, fasten with the melted mixture, a thin strip of wood, about one inch wide; let it project above the top of the box about an inch, and be kept about half an inch from the bottom; this answers as a spout for pouring the honey into the feeder, and when not in use, it should be stopped up. Have for the lid of the box, a piece of glass with the corner cut off next the spout, so as to cover the feeder and keep the bees in, and at the same time allow the bee-keeper to see when they have consumed all their food. The feeder is now complete, with one important exception; it has, as yet no way of admitting the bees. On the outside corners of one of the ends, glue or tack two strips, inch and a half wide, extending down to the bottom of the box, and half an inch from the top; fasten over them a piece of thin board, (paste-board will answer.) You have now a shallow passage without top or bottom, outside of your feeder; give it a top of any kind; cut out just below the level of this top, a passage into the feeder for the bees. It is now complete, and when properly placed over any hole on the top of the hive, will admit the bees from the hive, into the shallow passage which has no bottom, and through this into the feeder. Such a feeder will not only be cheap, but it might almost be made by a child, and yet it will answer every purpose most admirably. If you have no wooden box that will answer, a feeder may be made of pasteboard, and if brushed with the melted mixture it will be honey-tight. By packing cotton or wool around it, it might be used in most hives, even in the dead of Winter. Bees however, ought never to need feeding in Winter, and if they do, it will always be unsafe at this season to feed them with liquid honey.

I ought here to speak of the importance of *water* to the bees. It is absolutely indispensable when they are building comb, or raising brood. In the early Spring, they take advantage of the first warm weather, to bring it to their hives, and they may be seen busily drinking around pumps, drains, and other moist places. As they are not noticed frequenting such spots much, except in the early part of the season, many suppose that they need water only at this period. This is a great mistake, for they need it, and must have it, during the whole breeding season. But as soon as the grass starts, and the trees are covered with leaves, they prefer to sip the dew from them. If a few cold days come on, after the bees have commenced breeding, so as to prevent them from going abroad for water, a very serious check will be given to their operations. Even when it is not so cold as to prevent their leaving the hive, many become so chilled in their search for water, that they are not able to return.

Every wise bee-keeper will see that his bees have an abundant supply of water. If he has not some warm and sunny spot where they can safely obtain it, he will furnish them with shallow wooden troughs or vessels filled with pebbles, from which they can drink, without any risk of drowning, and where they will be sheltered from cold winds, and warmed by the genial rays of the sun. I believe that the reason why bees very much prefer the impure water of barn-yards and drains, is not because they find any medicinal quality in it, but because as it is *near* their hives and *warm*, they can fill themselves without being fatally chilled.

I have used water feeders of the same construction with my honey feeders, with great success. The bees are able to enter them at all times, as they are filled with the warm air of the hive, and thus breeding goes on, without interruption, and the lives of many bees are saved.

The same end may be obtained, by pouring daily, a few table spoonsfull of water into the hive, through one of the holes leading to the spare honey boxes. As soon as the weather becomes warm, and the bees can supply themselves from the dew on the grass and leaves, it will not be worth while to give them water in their hives.

When supplied with water in their hives, I advise that enough honey or sugar be added to it, to make it tolerably sweet. They will take it with greater relish, and it will stimulate them more powerfully to the raising of brood.

I come now to mention a substitute for liquid honey, the value of which has been extensively and thoroughly tested in Germany, and which I have used with great advantage. It was not discovered by Dzierzon, although he speaks of its excellence, in the most decided terms. The article to which I refer, is *plain sugar candy*, or as it is often called, barley candy. It has been ascertained that about four pounds of this, will sustain a colony during the Winter, when they have scarcely any honey in their hive! If it is placed where they can get access to it without being chilled, they will cluster upon it, and gradually eat it up. It not only goes further than double the quantity of liquid honey which could be bought for the same money, but is found to agree with the bees perfectly; while the liquid honey is almost sure to sour in the unsealed cells, and expose them to dangerous, and often fatal attacks of dysentery. I have sometimes, in the old-fashioned box hives, pushed sticks of candy between the ranges of comb, and have found it even then to answer a good purpose. In any hive which has surplus honey boxes, the candy may be put into a small box, which after being covered thoroughly with cotton or wool, may have another box put over it, the outside of which may be also covered. Unless great precautions are used, the boxes will be so cold, that the bees will not be able to enter them in Winter, and may thus perish in close proximity to abundant stores.

In my hives, the candy may be laid on the top of the frames, in the shallow chamber between the frames and the honey-board; it will here, if the honey-board is covered with straw, be always accessible to the bees, even in the coldest weather. I sometimes put it directly into a frame, and confine it with a piece of twine, or fine wire.

I have made a very convenient use of sugar candy, as a bee-feed in the Summer, when I wished to give small colonies a little food, and yet not to be at the trouble to use a feeder, or incur the risk of their being robbed by putting it where strange bees might be attracted by the scent. A small stick of candy, slid in on the bottom-board, under the frames, answers admirably for such a purpose. If a little liquid food must be used in warm weather, I advise that it be the best white sugar, dissolved in water; this makes an admirable food; costs but little more than brown sugar, and has no smell to tempt robbers to try to gain an entrance into the hive.

If the Apiarian is skillful, and attends to his bees, at the proper time, they will rarely need much feeding; if he manages them in such a manner that this is frequently and extensively needed, I can assure him, if he has not already found it out to his sorrow, that his bees will be nothing but a bill of cost and vexation.

The question how much honey a colony of bees needs, in order to carry them safely through the perils of Winter, is one to which it is impossible to give an answer which will be definite, under all circumstances. Very much will depend upon the hive in which they are kept, and the forwardness of the ensuing Spring; (see Chapter on Protection.) It is often absolutely impossible in the common hives, to form any reliable estimate, as to the quantity of honey which they contain, for the combs are often so heavy with bee-bread, as entirely to deceive even the most experienced bee-keeper.

I should always wish to leave at least 20 lbs. of honey in a hive; and as I can examine each comb, I am never at a loss to know how much a colony has. If I have the least apprehension that their supplies may fail, I prefer to put a few pounds of sugar candy where they can easily get access to it, in case of need. In my hive, the careful bee-keeper may not only know the exact extent of the resources of each hive, in the Fall, but he may, very early in the Spring, ascertain precisely how much honey is still on hand, and whether his bees need feeding, in order to preserve their lives. It is a shameful fact that a large number of colonies perish after they have begun to fly out, and when, they might easily have been saved, in any kind of hive.

Feeding, to make a profit by selling the Honey stored up by the Bees.

For many years, Apiarians have attempted to make the feeding of bees on a large scale, profitable to their owners. All such attempts however, must, from the very nature of the case, meet with very limited success. If large quantities of cheap West India honey are fed to the bees in the Fall, they are induced to fill their hives to such an extent, that in the Spring, the queen does not find the necessary accommodations for breeding. If they are largely fed in the Spring, the case is still worse; It must therefore be obvious that the feeding of cheap honey can only be made profitable where it serves as a substitute for an equal quantity of choice honey taken from the bees. In the latter part of Summer, the Apiarian may take away from the main hive, some of the combs which contain the best honey, and replace them with combs into which he has poured the cheaper article; or if he has no spare combs on hand, he may slice off the covers of the cells, drain out the honey, fill the empty combs with West India honey, and return them to the bees: giving them at the same time, the additional food which they need to elaborate wax to seal them over. If he attempts to take away their full combs, and gives them honey in order to enable them, first to replace their combs, and then to fill them, the operation, will result in a loss, instead of a gain.

I am aware that for a number of years, persons have attempted to derive a profit from supplying the markets of some of our large cities, with an article professing to be the best of honey, but which has been nothing more than the cheap West India honey fed to the bees, and stored up by them in new comb. In the City of Philadelphia, large quantities of such honey have been sold at the highest prices, and *perhaps* at some profit to the persons who have fed it to their bees. Within the last two years, however, the article has become so well known that it can hardly be sold at any price; as those who purchase honey, instead of paying 25 cents per pound for West India honey in the comb, much prefer to buy it, (if they want it at all,) for 6 or 7 cents, in a liquid state! It must be perfectly obvious that to sell a cheap and ill-flavored article at a high price, under the pretence that it is a superior article, is nothing less than downright cheating.

I am perfectly well aware that many persons imagine that if any thing *sweet* is fed to bees, they will quickly transmute it into the purest nectar. There is, however, no more truth in such a conceit, than there would be in that of a man who supposed that he had found the veritable philosopher's stone; and that he was able to change all our copper and silver coins into the purest gold! Bees to be sure, can make white and beautiful *comb*, from almost any kind of sweet; and why? because wax is a natural secretion of the bee, and can be made from any sweet; just as fat can be put upon the ribs of an ox, by any kind of nourishing food.

"But," some of my readers may ask, "do you mean to assert that bees do not secrete honey out of the raw material which they gather, or which is furnished to them, just as cows secrete milk from grass and hay?" I certainly do mean to assert that they can do nothing of the kind, and no intelligent man who has carefully *studied their habits*, will for a moment, venture to affirm that they can, unless for the sake of "filthy lucre," he is attempting to deceive an unwary community. What bee-keeper does not know, or rather ought not to know that the quality of honey depends entirely upon the sources from whence it is gathered; and that the different kinds of honey can easily be distinguished by any one who is a judge of the article.

Apple-blossom honey, white clover honey, buckwheat honey, and all the different kinds of honey, each has its own peculiar flavor, and it is utterly amazing how any sensible man, acquainted with bees, can be so deluded as to imagine any thing to the contrary. But as this is a matter of great practical importance, let us examine it more closely.

When bees are engaged in rapidly storing up honey in their combs, they may be seen, as *soon* as they return from the fields, or from the feeding boxes, putting their heads at once into the cells, and disgorging the contents of their "honey-bags." Now that the contents of their sacs undergo no change at all, during the short time that they remain in them, I will not absolutely affirm, because I have endeavored, through this whole treatise, never to assert positively when I had not positive evidence for so doing: but that they can undergo but a *very slight* change, must be evident from the fact that when thus stored up, the different kinds of honey or sugar can be almost if not quite as readily distinguished as before they were fed to the bees. The only perceptible change which they appear to undergo in the cells, is to have the large quantity of water evaporated from them, which is added from thoughtlessness, or from the vain expectation that it will be just so much water sold for honey, to the defrauded purchaser! This evaporation of the water from the honey by the heat of the hive, is about the only marked change that it appears to undergo, from its natural state in the nectaries of the blossoms; and it is exceedingly interesting to see how unwilling bees are to seal up honey, until it is reduced to such a consistency that there is no danger of its souring in the cells. They are as careful as to the quality of their nectar, as the good lady of the house is, to have the syrup of her preserves boiled down to a suitable thickness to keep them sweet.

Let all who for any purpose whatever, feed bees, keep this fact in mind, and never add to the food which they give them, more water than is absolutely necessary. To do so, is a piece of as great stupidity as to pour a barrel of water into the sugar pans, for every barrel of sap from the maples, or juice from the canes! If a strong colony is set upon a platform scale, it will be found on a pleasant day, during the height of the honey harvest, to gain a number of pounds; if examined again, early next morning, it will be found to have lost considerably, during the night. This is owing to the evaporation of the water from the freshly gathered honey, and it may often be seen running down in quite a stream from the bottom-board.

Those who feed cheap honey to sell it in the market at a high advance over its first cost, are either deceivers or deceived; if any of my readers have been deceived by the plausible representations of ignorant or unprincipled men, I trust they will be able from these remarks, to see exactly *how* they have been deceived, and they will no longer persist in an adulteration, the profits of which can never be great, and the morality of which can never be defended. A man who offers for sale, inferior honey, or sugar which he calls honey, and which he is able to sell because it is stored in white comb, to those who would never purchase it if they knew what it was, or once had a taste of it, is not a whit more honest, if he understands the nature of the article in which he deals, than a person engaged in counterfeiting the current coin of the realm: for poor honey in white comb, is no less a fraud than eagles or dollars, golden to be sure, on their honest exteriors, but containing a baser metal within! "The Golden Age" of bee-keeping, in which inferior honey can be quickly transmuted into such balmy spoils as are gathered by the bees of Hybla, has not yet dawned upon us; or at least only in the fairy visions of the poet who saw "A golden hive, on a Golden Bank, Where golden bees, by alchemical prank, Gathered Gold instead of Honey."

If a pound of West India honey costs about 6 cents, and the bees use, as they will, about one pound to make the comb in which it is stored, it costs the producer at least 12 cents a pound, and if to this, he adds, say 5 cents more, for extra time and labor in feeding, then his inferior honey costs him at least as much as the market price of the very best honey on the spot where it is produced! If the bee-keeper allows his bees to make what they will, from the blossoms, and then begins to feed, after he has harvested the produce from the natural supplies, the advance over the first cost will hardly pay for the trouble, even if it were fair to palm off such inferior honey as a first-rate article. If, however, bees are fed on this food very largely in the latter part of Summer, they will fill up their hive with it, before they put it into the spare honey boxes, and the production of brood will often be most seriously interfered with, at a season of the year when it is important to have the hives well stocked with bees, that they may winter to the best advantage.

If Apiarians are anxious to have large quantities of choice honey, let them manage their bees so as to have powerful stocks in the early Spring, and they will then be able to have heavy purses and light consciences into the bargain. I shall now show how liquid honey, exceedingly beautiful to the eye, and tempting to the taste, may be made to great advantage.

Dissolve two pounds of the purest white sugar, in as much hot water as will be just necessary to reduce it to a syrup; take one pound of the nicest white clover honey, (any other light colored honey of good flavor will answer,) and after warming it, add it to the sugar syrup, and stir the contents. When cool, this compound will be pronounced, even by the best judges of honey, to be one of the most luscious articles which they ever tasted; and will be, by almost every one, preferred to the unmixed honey. Refined loaf sugar is a perfectly pure and inodorous sweet, and one pound of honey will communicate the honey flavor, in high perfection, to twice that quantity of sugar: while the new article will be destitute of that smarting taste which honey alone, so often has, and will be often found to agree perfectly with those who cannot eat the clear honey with impunity. If those engaged in the artificial manufacture of honey, never brought any thing worse than this, to the market, the purchasers would have no reason to complain. As however, the compound can be furnished much cheaper than the pure honey, many may prefer to purchase the materials, and mix them themselves. If desired, any kind of flavor may be given to the manufactured article; thus it may be made to resemble in fragrance, the classic honey of Mount Hymettus, by adding to it the fine aroma of the lemon balm, or wild thyme; or it may have the flavor of the orange groves, or the delicate fragrance of beds of roses washed with dew.

I have recently ascertained that if two pounds of the best refined sugar be added to one of common maple sugar, the compound will be a light colored article, retaining perfectly the maple taste, and yet far superior to the common maple sugar. After making this discovery, I learned that a large part of the very nicest maple sugar is made in this way!

Attempts have been made to feed to bees, to be stored in the honey boxes, a mixture of the whitest honey and loaf sugar; but the result shows a loss rather than a gain. The mixture, before it is fed, will cost about 10 cents per pound. At the very furthest, not more than one half of what is fed, can be secured in the comb, for it requires about one pound of honey, to manufacture comb enough to hold a pound of honey. The actual cost of the honey in the comb, will therefore be, at least 20 cents per pound; and the pure white clover honey can be bought for less than that. Those who desire to have something exceedingly beautiful to the eye, and delicate to the taste, at a season when the bees are not storing up honey from the blossoms, and in situations where the natural supply is of an inferior quality, if they do not regard expense, can place upon their tables, something which will be pronounced by the best judges, a little superior to any thing they ever tasted before.

I have repeatedly spoken of the great care which is necessary to prevent bees from getting a taste of forbidden sweets, so as to be tempted to engage in dishonest courses. The experienced Apiarian will fully appreciate the necessity of these cautions, and the inexperienced, if they neglect them, will be taught a lesson that they will not soon forget. Let it be remembered that the bee was intended to gather its sweets from the nectaries of flowers: to use the exquisitely beautiful language of him whose wonderful writings supply us on almost every subject, with the richest thoughts and happiest illustrations, they were created to

"Make boot upon the Summer's velvet buds, Which pillage they with merry march bring home To the tent royal of their emperor: Who, busied in his majesty, surveys The singing masons, building roofs of gold."—*Shakspeare.*

When thus engaged, the bees work in perfect accordance with their natural instincts, and seem to have little or no disposition to meddle with property that does not belong to them. If however, their incautious owner tempts them with liquid food, especially at times when they can obtain nothing from the blossoms, they seem to be so infatuated with such easy gatherings, as to lose all discretion, and they will perish by thousands, if the vessels which contain the food are not furnished with floats, on which they can stand and help themselves in safety.

The fly was intended to feed, not upon the blossoms, but upon food in which, without care, it could easily be drowned; and hence it alights most cautiously, on the edge of any vessel containing liquid food, and warily helps itself: while the poor bee, without any caution, plunges

right in and speedily perishes. The sad fate of their unfortunate companions, does not in the least, deter others who approach the tempting lure:
but they madly alight on the bodies of the dying and the dead, to share
the same miserable end! No one can understand the full extent of their
infatuation, until after seeing a confectioner's shop, assailed by thousands
and tens of thousands of hungry bees. I have seen thousands strained out
from the syrups in which they had perished; thousands more alighting
even upon the boiling sweets; the floors covered, and windows darkened
with bees, some crawling, others flying, and others still, so completely
daubed as to be able neither to crawl nor fly; not one bee in ten able to
carry home its ill-gotten spoils, and yet the air filled with new hosts of
thoughtless comers.

It will be for the interest of all engaged in the manufacture of
candy and syrups, to fit gauze wire windows and doors to their premises, and thus save themselves from constant loss and annoyance: for if
only one bee in a hundred escapes with his load, the confectioner will
be subjected in the course of the season to serious loss. I once furnished
such an establishment, after the bees had commenced their depredations, with such protection; and when they found themselves excluded,
they lit on the wire by thousands, and fairly squealed with vexation and
disappointment, as they tried to force a passage through the meshes. At
last as they were daring enough to descend the chimney, reeking with
sweet odors, even although the most who attempted it, fell with scorched
wings into the fire, it became necessary to put wire gauze over the top of
the chimney also!

How often, as I have seen thousands of bees, in such places
destroyed, and thousands more deprived of all ability to fly, and hopelessly struggling in the deluding sweets, and yet thousands more blindly
hovering over them, all unmindful of their danger, and apparently eager
to share the same destruction, how often has the spectacle of their infatuation seemed to me, to be an exact picture of the woful delusion of those
who surrender themselves to the fatal influences of the intoxicating cup.
Even although they see the miserable victims of this degrading vice, falling
all around them, into premature and dishonored graves, they still press
on, madly trampling as it were, over their dead and dying bodies, that
they too may sink into the same abyss of agonies, and that their sun may
also go down in darkness and hopeless gloom. Even although they know
that the next cup may send them, with all their sins upon their heads,

to the dread tribunal of their God, that cup of bitter sorrows and untold degradation, they will drain even to its most loathsome dregs.

The avaricious bee that despised the slow process of extracting nectar from "every opening flower," and plunged recklessly into the tempting sweets, has ample time to bewail its folly. Even if it has not paid the forfeit of its life, but has been able to obtain its fill, it returns home with all its beautiful plumage sullied and besmeared, and with a woe-begone look, and sorrowful note, in marked contrast with the bright hues and merry sounds with which the industrious bee returns from its happy rovings amid "the budding honey flowers, and sweetly breathing fields."

Just so, has many a pilgrim from the golden shores of California and Australia, returned; enfeebled in body and mind, bankrupt often in character and happiness, if not in purse, and unfitted in every way, for the calm and sober pursuits of common industry; while thousands, yes, and tens of thousands too, shall never more behold their once happy homes. Bibles and Sabbaths, altars and firesides, parents and friends, wife and children, how often have all these been wantonly abandoned, in the accursed greed for gain, by those who might have been happy and prosperous at home, and who wandered from its sacred precincts only because they were determined to make the possession of wealth, the chief object of life, but whose bones now lie amid the coral reefs of the ocean, or moulder in the howling wastes of the "overland passage;" just as the bones of the unbelieving Israelites whitened the sands of the desert. Of those who have reached the "land of" golden "promise," how many have died in despair, or worse still, are living so besotted by vice, so lost to all power of virtuous resolutions, that they shall never more see the happy homes from which they so thoughtlessly wandered, never more hear the soft accents of loving friends; never more worship God, in a peaceful Sanctuary, or ever again behold an opened Bible!

"Gold! Gold! Gold! Gold! Bright and yellow, hard and cold, Molten, graven, hammer'd, and roll'd; Heavy to get, and light to hold; Hoarded, barter'd, bought, and sold, Stolen, borrow'd, squander'd, doled: Spurn'd by the young, but hugg'd by the old To the very verge of the churchyard mould; Price of many a crime untold; Gold! Gold! Gold! Gold! Good or bad a thousand-fold! How widely its agencies vary— To save—to ruin—to curse—to bless— As even its minted coins express, Now stamp'd with the image of Good Queen Bess, And now of a Bloody Mary!" *Hood.*

CHAPTER XVII. HONEY. PASTURAGE. OVERSTOCKING.

In the chapter on Feeding, it has already been stated that honey is not a natural secretion of the bee, but a substance obtained from the nectaries of the blossoms; it is not therefore, made, but merely gathered by the bees. The truth is well expressed in the lines so familiar to most of us from our childhood,

"How doth the little busy bee Improve each shining hour, And *gather* honey all the day From every opening flower."

Bees not only gather honey from the blossoms, but often obtain it in large quantities from what have been called honey dews; "a term applied to those sweet, clammy drops that glitter on the foliage of many trees in hot weather." Two different opinions have been zealously advocated as to the origin of honey-dews. By some, they are considered a natural exudation from the leaves of trees, a perspiration as it were, occasioned often by ill health, though sometimes a provision to enable the plants to resist the fervent heats to which they are exposed. Others insist that this sweet substance is discharged from the bodies of those aphides or small lice which infest the leaves of so many plants. Unquestionably they are produced in both ways.

Messrs. Kirby and Spence, in their interesting work on Entomology, have given a description of the kind of honey-dew furnished by the aphides.

"The loves of the ants and the aphides have long been celebrated; and that there is a connection between them, you may, at any time in the proper season, convince yourself; for you will always find the former very busy on those trees and plants on which the latter abound; and if you examine more closely, you will discover that the object of the ants, in thus attending upon the aphides, is to obtain the saccharine fluid secreted by them, which may well be denominated their milk. This fluid, which is scarcely inferior to honey in sweetness, issues in limpid drops from the abdomen of these insects, not only by the ordinary passage, but also by two setiform tubes placed, one on each side, just above it. Their sucker being inserted in the tender bark, is without intermission employed in absorbing the sap, which, after it has passed through their system, they keep continually discharging by these organs. When no ants attend them, by a certain jerk of the body, which takes place at regular intervals, they ejaculate it to a distance."

"Mr. Knight once observed," says Bevan, "a shower of honey-dew descending in innumerable small globules, near one of his oak-trees, *on the 1st of September*; he cut off one of the branches, took it into the house, and holding it in a stream of light, which was purposely admitted through a small opening, distinctly saw the aphides ejecting the fluid from their bodies with considerable force, and this accounts for its being frequently found in situations where it could not have arrived by the mere influence of gravitation. The drops that are thus spurted out, unless interrupted by the surrounding foliage, or some other interposing body, fall upon the ground; and the spots may often be observed, for some time, beneath and around the trees affected with honey-dew, till washed away by the rain. The power which these insects possess of ejecting the fluid from their bodies, seems to have been wisely instituted to preserve cleanliness in each individual fly, and indeed for the preservation of the whole family; for pressing as they do upon one another, they would otherwise soon be glued together, and rendered incapable of stirring. On looking steadfastly at a group of these insects (*Aphides Salicis*) while feeding on the bark of the willow, their superior size enables us to perceive some of them elevating their bodies and emitting a transparent substance in the form of a small shower."

"Nor scorn ye now, fond elves, the foliage sear, When the light aphids, arm'd with puny spear, Probe each emulgent vein, till bright below, Like falling stars, clear drops of nectar glow." *Evans.*

"The *willow* accommodates the bees in a kind of threefold succession; from the flowers they obtain both honey and farina;—from the bark propolis;—and the leaves frequently afford them honey-dew at a time when other resources are beginning to fail."

"Honey-dew usually appears upon the leaves as a viscid, transparent substance, as sweet as honey itself, sometimes in the form of globules, at others resembling a syrup; it is generally most abundant from the middle of June to the middle of July, sometimes as late as September."

"It is found chiefly upon the *oak*, the *elm*, the *maple*, the *plane*, the *sycamore*, the *lime*, the *hazel*, and the *blackberry*; occasionally also on the *cherry*, *currant*, and other fruit trees. Sometimes only one species of trees is affected at a time. The oak generally affords the largest quantity. At the season of its greatest abundance, the happy humming noise of the bees may be heard at a considerable distance from the trees, sometimes nearly equalling in loudness the united hum of swarming."

In some seasons, extraordinary quantities of honey are furnished by the honey-dews, and bees will often, in a few days, fill their hives with it. If at such times, they can be furnished with empty combs, the amount stored up by them, will be truly wonderful. No certain reliance, however, can be placed upon this article of bee-food, as in some years, there is scarcely any to be found, and it is only once in three or four years, that it is very abundant. The honey obtained from this source, is generally of a very good quality, though seldom as clear as that gathered from the choicest blossoms.

The quality of honey is exceedingly various, some being dark, and often bitter and disagreeable to the taste, while occasionally it is gathered from poisonous flowers, and is very noxious to the human system.

An intelligent Mandingo African informed a lady of my acquaintance, that they do not in his country, dare to eat *unsealed* honey, until it is first *boiled*. In some of the Southern States, all unsealed honey is generally rejected. It appears to me highly probable that the noxious qualities of the honey gathered from some flowers, is, for the most part, evaporated, before it is sealed over by the bees, while the honey is thickening in the cells. Boiling the honey, would, of course, expel it much more effectually, and it is a well ascertained fact that some persons are not able to eat even the best honey with impunity, until after it has been boiled! I believe that if persons who are injured by honey would subject it to this operation, they would usually find it to exert no injurious influence on the system. Honey is improved by age, and many are able to use with impunity, that which has been for a long time, in the hive, and which seems to be much milder than any freshly gathered by the bees.

Honey, when taken from the bees, should be carefully put where it will be safe from all intruders, and where it will not be exposed to so low a temperature as to candy in the cells. The little red ant, and the large black ant are extravagantly fond of it, and unless placed where they cannot reach it, they will soon carry off large quantities. I paste paper over all my boxes, glasses, &c., so as to make them air-tight, and carefully store them away for future use. If it is drained from the combs, it may be kept in tight vessels, although in this state it will be almost sure to candy. By putting the vessels in water, and bringing it to the boiling point, it will be as nice as when first strained from the comb. In this way, I prefer to keep the larger portion of my honey. The appearance of white honey in the comb, is however, so beautiful, that many will prefer to keep it in this form, especially, if intended for sale.

In my hives, it may be taken from the bees, in a great variety of ways. Some may prefer to construct the main hive in such a form, that the surplus honey can be taken from it, on the frames. Others will prefer to take it on frames put in an upper box; Glass vessels of almost any size or form will make beautiful receptacles for the spare honey. They ought always, however, to have a piece of comb fastened in them, before they are given to the bees; and if the weather is cool, they must be carefully covered with something warm, or they will part with their heat so quickly, as to discourage the bees from building in them. Unless warmly covered, glass vessels will often be so lined with moisture, as to annoy the bees. This is occasioned by the rapid evaporation of the water from the newly gathered honey, All hives during the height of the gathering season, abound in moisture, and this no doubt furnishes the bees, for the most part, with the water they then need.

Honey, when stored in a pint tumbler, just large enough to receive one comb, has a most beautiful appearance, and may be easily taken out whole, and placed in an elegant shape upon the table. The expense of such glass vessels is one objection to their use; the ease with which they part with their heat, another, and a more serious objection still, is the fact that the shallow cells, so many of which must be made in a round vessel, require as large a consumption of honey for their wax covers, as those which hold more than twice their quantity of honey.

I prefer rectangular boxes made of pasteboard, to any other: they are neat, warm and cheap; and if a small piece of glass is pasted in one of their ends, the Apiarian can always see when they are full. When the honey is taken from the bees, the box has its cover put on, and is pasted tight, so as to exclude air and insects. In this form, honey may be packed, and sent to market very conveniently: and when the boxes are opened, the purchaser can always see the quality of the article which he buys. The box in which these small boxes of honey are packed in order to be sent to market, should be furnished with rope handles, so that it can be easily lifted, without the least jarring. Honey should be handled with just as much care as glass. A box, four inches wide, will admit of two combs, and if small pieces of comb are put in the top, the bees will build them, of the proper dimensions, and will thus make them too large for brood combs, and of the best size to contain their surplus honey. The use of my hives enables the Apiarian to get access to all the comb which he needs for such purposes, and he will find it to his interest, never to give the bees a box which does not contain some comb, as well for encouragement as for a pattern. I have never seen the use of pasteboard boxes suggested, but after experimenting with a great many materials, I believe they will be found, all things considered, preferable to any others. Wooden boxes, with a piece of glass, are very good for storing honey: but they are much more expensive than those made of pasteboard, and the covers cannot be removed so conveniently.

Honey may be safely removed from the surplus honey boxes of my hives, even by the most timid. When the outside case which covers the boxes, is elevated, a shield is thrown between the Apiarian and the bees which are entering and leaving the hive. Before removing a vessel or box, a thin knife should be carefully passed under it, so as to loosen the attachments of the comb to the honey-board, without injuring the bees; then a small piece of tin or zinc may be pushed under to prevent the bees that are below, from coming up, when the honey is removed. The Apiarian should now tap gently on the box, and the bees in it, perceiving that they are separated from the main hive, will at once proceed to fill themselves, so as to save as much as possible, of their precious sweets. In about five minutes, or as soon as they are full, and run over the combs, trying to get out, the glass or box may at once be removed, and they will fly directly to the hive with what they have been able to secure. Bees under such circumstances, *never* attempt to sting, and a child of ten years, may remove, with ease and safety, all their surplus stores. If a person is too timid to approach a hive when any bees are flying, the honey may be removed towards evening, or early in the morning, before the bees are flying, in any considerable numbers. In performing this operation, it should always be borne in mind, that large quantities of honey should never be taken from them at once, unless when the honey-harvest is over. Bees are exceedingly discouraged by such wholesale appropriations, and often refuse entirely, to work in the empty boxes, even although honey abounds in the fields. Not infrequently when large boxes are removed, and being found only partially filled, are returned, the bees will carry every particle of honey down into the main hive! If, however, the honey is removed in small boxes, one at a time, and an empty box with guide comb is put instantly in its place, the bees, so far from being discouraged, work with more than their wonted energy, and usually begin in a few hours, to enlarge the comb.

I would here repeat the caution already given, against needlessly opening and shutting the hives, or in any way meddling with the bees so as to make them feel insecure in their possessions. Such a course tends to discourage them, and may seriously diminish the yield of honey.

If the Apiarian wishes to remove honey from the interior of the hive, he must remove the combs, as directed on page 195, and shake the bees off, on the alighting board, or directly into the hive.

PASTURAGE.

Some blossoms yield only pollen, and others only honey; but by far the largest number, both honey and pollen. Since the discovery that rye flour will answer so admirably as a substitute, before the bees are able to gather the pollen from the flowers, early blossoms producing pollen alone, are not so important in the vicinity of an Apiary. Willows are among the most desirable trees to have within reach of the bees: some kinds of willow put out their catkins very early, and yield an abundance of both bee-bread and honey. All the willows furnish an abundance of food for the bees; and as there is considerable difference in the time of their blossoming, it is desirable to have such varieties as will furnish the bees with food, as long as possible.

The Sugar Maple furnishes a large supply of very delicious honey, and its blossoms hanging in drooping fringes, will be all alive with bees. The Apricot, Peach, Plum and Cherry are much frequented by the bees; Pears and Apples furnish very copious supplies of the richest honey. The Tulip tree, *Liriodendron*, is probably one of the greatest honey-producing trees in the world. In rich lands this magnificent tree will grow over one hundred feet high, and when covered with its large bell-shaped blossoms of mingled green and golden yellow, it is one of the most beautiful trees in the world. The blossoms are expanding in succession, often for more than two weeks, and a new swarm will frequently fill its hive from these trees alone. The honey though dark in color, is of a rich flavor. This tree has been successfully cultivated as a shade tree, even as far North as Southern Vermont, and for the extraordinary beauty of its foliage and blossoms, deserves to be introduced wherever it can be made to grow. The Winter of 1851-2, was exceedingly cold, the thermometer in Greenfield, Mass. sinking as low as 30° below zero, and yet a tulip tree not only survived the Winter uninjured, but was covered the following season with blossoms.

The American Linden or Bass Wood, is another tree which yields large supplies of very pure and white honey. It is one of our most beautiful native trees, and ought to be planted much more extensively than it is, in our villages and country seats. The English Linden is worthless for bees, and in many places, has been so infested by worms, as to make it necessary to cut it down.

The Linden blossoms soon after the white clover begins to fail, and a majestic tree covered with its yellow clusters, at a season when very few blossoms are to be seen, is a sight most beautiful and refreshing.

"Here their delicious task, the fervent bees In swarming millions tend: around, athwart, Through the soft air the busy nations fly, Cling to the bud, and with inserted tube, Suck its pure essence, its etherial soul." *Thomson.*

Our villages would be much more attractive, if instead of being filled as they often are, almost exclusively with maples and elms, they were adorned with a greater variety of our native trees. The remark has often been made, that these trees are much more highly valued abroad than at home, and that to see them in perfection, we must either visit their native forests, or the pleasure grounds of some wealthy English or European gentleman.

Of all the various sources from which the bees derive their supplies, white clover is the most important. It yields large quantities of very white honey, and of the purest quality, and wherever it flourishes in abundance, the honey-bee will always gather a rich harvest. In this country at least, it seems to be the most certain reliance of the Apiary. It blossoms at a season of the year when the weather is usually both dry and hot, and the bees gather the honey from it, after the sun has dried off the dew: so that its juices are very thick, and almost ready to be sealed over at once in the cells.

Every observant bee-keeper must have noticed, that in some seasons, the blossoms of various kinds yield much less honey than in others. Perhaps no plant varies so little in this respect, as the white clover. This clover ought to be much more extensively cultivated than it now is, and I consider myself as conferring a benefit not only on bee-keepers, but on the agricultural community at large, in being able to state on the authority of one of New England's ablest practical farmers, and writers on agricultural subjects, Hon. Frederick Holbrook, of Brattleboro', Vermont, that the common white clover may be cultivated on some soils to very great profit, as a hay crop. In an article for the New England Farmer, for May, 1853, he speaks as follows:—

"The more general sowing of white clover-seed is confidently recommended. If land is in good heart at the time of stocking it to grass, white clover sown with the other grass-seeds will thicken up the bottom of mowings, growing some eight or ten inches high and in a thick mat, and the burden of hay will prove much heavier than it seemed likely to be before mowing. Soon after the practice of sowing white clover on the tillage-fields commences, the plant will begin to show itself in various places on the farm, and ultimately gets pretty well scattered over the pastures, as it seeds very profusely, and the seeds are carried from place to place in the manure and otherwise. The price of the seed per pound in market is high; but then one pound of it will seed more land, than two pounds of red clover seed; so that in fact the former is the cheaper seed of the two, for an acre."

"Red-top, red clover and white clover seeds, sown together, produce a quality of hay universally relished by stock. My practice is, to seed all dry, sandy and gravelly lands with this mixture. The red and white clover pretty much make the crop the first year; the second year, the red clover begins to disappear, and the red-top to take its place; and after that, the red-top and white clover have full possession and make the very best hay for horses or oxen, milch cows or young stock, that I have been able to produce. The crop per acre, as compared with herds-grass, is not so bulky; but tested by weight and by spending quality in the Winter, it is much the most valuable."

"Herds-grass hay grown on moist uplands or reclaimed meadows, and swamps of a mucky soil, or lands not overcharged with silica, is of good quality; but when grown on sandy and gravelly soils abounding in silex, the stalks are hard, wiry, coated with silicates as with glass, and neither horses nor cattle will eat it as well, or thrive as well on it as on hay made of red-top and clover; and as for milch cows, they winter badly on it, and do not give out the milk as when fed on softer and more succulent hay."

By managing white clover, according to Mr. Holbrook's plan, it might be made to blossom abundantly in the second crop, and thus lengthen out, to very great advantage, the pasture for the bees. For fear that any of my readers might suspect Mr. Holbrook of looking at the white clover, through a pair of *bee-spectacles*, I would add that although he has ten acres of it in mowing, he has no bees, and has never particularly interested himself in this branch of rural economy. When we can succeed in directing the attention of such men to bee-culture, we may hope to see as rapid an advance in this as in some other important branches of agriculture.

Sweet-scented clover, (*Mellilotus Leucantha*,) affords a rich bee-pasturage. It blossoms the second year from the seed, and grows to a great height, and is always swarming with bees until quite late in the Fall. Attempts have been made to cultivate it for the sake of its value as a hay crop, but it has been found too coarse in its texture, to be very profitable. Where many bees are kept, it might however, be so valuable for them as to justify its extensive cultivation. During the early part of the season, it might be mowed and fed to the cattle, in a green and tender state, and allowed to blossom later in the season, when the bees can find but few sources to gather from.

For years, I have attempted to procure, through botanists, a hybrid or cross between the red and white clover, in order to get something with the rich honey-producing properties of the red, and yet with a short blossom into which the honey-bee might insert its proboscis. The red clover produces a vast amount of food for the bumble-bee, but is of no use at all to the honey-bee. I had hoped to procure a variety which might answer all the purposes of our farmers as a field crop. Quite recently I have ascertained that such a hybrid has been originated in Sweden, and has been imported into this country, by Mr. B. C. Rogers, of Philadelphia. It grows even taller than the red clover, bears many blossoms on a stalk which are small, resembling the white, and is said to be preferred by cattle, to any other kind of grass, while it answers admirably for bees.

Buckwheat furnishes a most excellent Fall feed for bees; the honey is not so well-flavored as some other kinds, but it comes at a season when it is highly important to the bees, and they are often able to fill their hives with a generous supply against Winter. Buckwheat honey is gathered when the dew is upon the blossoms, and instead of being thick, like white clover honey, is often quite thin; the bees sweat out a large portion of its moisture, but still they do not exhaust the whole of it, and in wet seasons especially, it is liable to sour in the cells. Honey gathered in a dry season, is always thicker, and of course more valuable than that gathered in a wet one, as it contains much less water. Buckwheat is uncertain in its honey-bearing qualities; in some seasons, it yields next to none, and hardly a bee will be seen upon a large field, while in others, it furnishes an extraordinary supply. The most practical and scientific agriculturists agree that so far from being an impoverishing crop, it is on many soils, one of the most profitable that can be raised. Every bee-keeper should have some in the vicinity of his hives.

The raspberry, it is well known, is a great favorite with the bees; and the honey supplied by it, is very delicious. Those parts of New England, which are hilly and rough, are often covered with the wild raspberry, and would furnish food for numerous colonies of bees.

It will be observed that thus far, I have said nothing about cultivating flowers in the garden, to supply the bees with food. What can be done in this way, is of scarcely any account; and it would be almost as reasonable to expect to furnish food for a stock of cattle, from a small grass plat, as honey for bees, from garden plants. The cultivation of bee-flowers is more a matter of pleasure than profit, to those who like to hear the happy hum of the busy bees, as they walk in their gardens. It hardly seems expedient, at least for the present, to cultivate any field crops except such as are profitable in themselves, without any reference to the bees.

Mignonnette is excellent for bees, but of all flowers, none seems to equal the Borage. It blossoms in June, and continues in bloom until severe frost, and is always covered with bees, even in dull weather, as its pendant blossoms keep the honey from the moisture; the honey yielded by it, is of a very superior quality. If any plant which does not in itself make a valuable crop, would justify cultivation, there is no doubt that borage would. An acre of it would support a large number of stocks. If in a village those who keep bees would unite together and secure the sowing of an acre, in their immediate vicinity, each person paying in proportion to the number of stocks kept, it might be found profitable. The plants should have about two feet of space every way, and after they covered the ground, would need no further attention. They would come into full blossom, cultivated in this manner, about the time that the white clover begins to fail, and would not only furnish rich pasture for the bees, but would keep them from the groceries and shops in which so many perish.

If those who are engaged in adorning our villages and country residences with shade trees, would be careful to set out a liberal allowance of such kinds as are not only beautiful to us, but attractive to the bees, in process of time the honey resources of the country might be very greatly increased.

OVERSTOCKING A DISTRICT WITH BEES.

I come now to a point of the very first importance to all interested in the cultivation of bees. If the opinions which the great majority of American bee-keepers entertain, are correct, then the keeping of bees must, in our country, be always an insignificant pursuit. I confess that I find it difficult to repress a smile, when the owner of a few hives, in a district where as many hundreds might be made to prosper, gravely imputes his ill success, to the fact that too many bees are kept in his vicinity! The truth is, that as bees are frequently managed, they are of but little value, even though in "a land flowing with milk and honey." If in the Spring, a colony of bees is prosperous and healthy, it will gather abundant stores, even if hundreds equally strong, are in its immediate vicinity, while if it is feeble, it will be of little or no value, even if there is not another swarm within a dozen miles of it.

Success in bee-keeping requires that a man should be in some things, a very close imitator of Napoleon, who always aimed to have an overwhelming force, at the right time and in the right place; so the bee-keeper must be sure that his colonies are numerous, just at the time when their numbers can be turned to the best account. If the bees cannot get up their numbers until the honey-harvest is well nigh gone, numbers will then be of as little service as many of the famous armies against which "the soldier of Europe" contended; which, after the fortunes of the campaign were decided, only served to swell the triumphant spoils of the mighty conqueror. A bee-keeper with feeble stocks in the Spring, which become strong only when there is nothing to get, is like a farmer who contrives to hire no hands to reap his harvests, but suffers the crops to rot upon the ground, and then at great expense, hires a number of stalworth laborers to idle about his premises and eat him out of house and home!

I do not believe that there is a *single square mile* in this whole country, which is overstocked with bees, unless it is one so unsuitable for bee-keeping as to make it unprofitable to attempt it at all. Such an assertion will doubtless, appear to many, very unguarded; and yet it is made advisedly, and I am happy to be able to confirm it, by reference to the experience of the largest cultivators in Europe. The following letter from Mr. Wagner, will I trust, do more than I can possibly do in any other way, to show our bee-keepers how mistaken they are in their opinion as to the danger of overstocking their districts, and also what large results might be obtained from a more extensive cultivation of bees.

York, March 16, 1853.

Dear Sir:

In reply to your enquiry respecting the *overstocking* of a district, I would say that the present opinion of the correspondents of the Bienenzeitung, appears to be that it *cannot readily be done*. Dzierzon says, in practice at least, "*it never is done;*" and Dr. Radlkofer, of Munich, the President of the second Apiarian Convention, declares that his apprehensions on that score were dissipated by observations which he had opportunity and occasion to make, when on his way home from the Convention. I have numerous accounts of Apiaries in pretty close proximity, containing from 200 to 300 colonies each. Ehrenfels had a thousand hives, at three separate establishments indeed, but so close to each other that he could visit them all in half an hour's ride; and he says that in 1801, the average net yield of his Apiaries was $2 per hive. In Russia and Hungary, Apiaries numbering from 2000 to 5000 colonies are said not to be unfrequent; and we know that as many as 4000 hives are oftentimes congregated, in Autumn, at one point on the heaths of Germany. Hence I think we need not fear that any district of this country, so distinguished for abundant natural vegetation and diversified culture, will very speedily be overstocked, particularly after the importance of having stocks populous early in the Spring, comes to be duly appreciated. A week or ten days of favorable weather, at that season, when pasturage abounds, will enable a *strong* colony to lay up an ample supply for the year, if its labor be properly directed.

Mr. Kaden, one of the ablest contributors to the Bienenzeitung, in the number for December, 1852, noticing the communication from Dr. Radlkofer, says: "I also concur in the opinion that a district of country cannot be overstocked with bees; and that, however numerous the colonies, all can procure sufficient sustenance if the surrounding country contain honey-yielding plants and vegetables, in the usual degree. Where utter barrenness prevails, the case is different, of course, as well as rare."

The Fifteenth Annual Meeting of German Agriculturists was held

in the City of Hanover, on the 10th of September, 1852, and in compli-
ance with the suggestions of the Apiarian Convention, a distinct section
devoted to bee-culture was instituted. The programme propounded
sixteen questions for discussion, the fourth of which was as follows:—

"Can a district of country embracing meadows, arable land,
orchards, and woodlands or forests, be so overstocked with bees, that
these may no longer find adequate sustenance and yield a remunerating
surplus of their products?"

This question was debated with considerable animation. The
Rev. Mr. Kleine, (nine-tenths of the correspondents of the Bee-Journal
are clergyman,) President of the section, gave it as his opinion that "it
was hardly conceivable that such a country could be overstocked with
bees." Counsellor Herwig, and the Rev. Mr. Wilkens, on the contrary,
maintained that "it might be overstocked." In reply, Assessor Heyne
remarked that "whatever might be supposed possible as an extreme case,
it was certain that as regards the kingdom of Hanover, it could not be even
remotely apprehended that too many Apiaries would ever be established;
and that consequently the greatest possible multiplication of colonies
might safely be aimed at and encouraged." At the same time, he advised
a proper distribution of Apiaries.

I might easily furnish you with more matter of this sort, and
designate a considerable number of Apiaries in various parts of Germany,
containing from 25 to 500 colonies. But the question would still recur,
do not these Apiaries occupy comparatively isolated positions? and at
this distance from the scene, it would obviously be impossible to give a
perfectly satisfactory answer.

According to the statistical tables of the kingdom of Hannover,
the annual production of bees-wax in the province of Lunenburg, is
300,000 lbs., about one half of which is exported; and assuming one pound
of wax as the yield of each hive, we must suppose that 300,000 hives are
annually "*brimstoned*" in the province; and assuming further, in view of
casualties, local influences, unfavorable seasons, &c., that only one-half
of the whole number of colonies maintained, produce a swarm each,
every year, it would require a total of at least 600,000 colonies, (141, to
each square mile,) to secure the result given in the tables.

The number of square miles stocked even to this extent, in this
country, are, I suspect, "few and far between." The Shakers at Lebanon,
have about 600 colonies; but I doubt whether a dozen Apiaries equally

large can be found in the Union. It is very evident, that this country is far from being overstocked; nor it is likely that it ever will be.

A German writer alleges that "the bees of Lunenburg, pay all the taxes assessed on their proprietors, and leave a surplus besides." The importance attached to bee-culture accounts in part for the remarkable fact that the people of a district so barren that it has been called "the Arabia of Germany," are almost without exception in easy and comfortable circumstances. Could not still more favorable results be obtained in this country under a rational system of management, availing itself of the aid of science, art and skill?

But, I am digressing. My design was to furnish you with an account of bee-culture as it exists *in an entire district of country*, in the hands of *the common peasantry*. This I thought would be more satisfactory, and convey a better idea of what may be done on a large scale, than any number of instances which might be selected of splendid success in isolated cases.

Very truly yours,

SAMUEL WAGNER.

Rev. L. L. Langstroth.

The question how far bees will fly in search of honey, has been very differently answered by different Apiarians. I am satisfied that they will fly over three miles in search of food, but I believe as a general rule, that if their food is not within a circle of about two miles in every direction from the Apiary, they will be able to store up but little surplus honey. The nearer, the better. In all my arrangements, I have made it a constant study to save *every step* for the bees that I possibly can, economizing to the very utmost, their time, which will all be transmuted into honey; an inspection of the Frontispiece of this treatise will exhibit the general aspect of the alighting board of my hives, and will show the intelligent Apiarian, with what ease bees will enter such a hive, even in very windy weather. By such arrangements, they will be able to store up more honey, even if they have to go a considerable distance in search of it, than they would in many other hives, when the honey abounded in their more immediate vicinity. Such considerations are entirely overlooked, by most bee-keepers, and they seem to imagine that they are matters of no importance. By the utter neglect of any kind of precautions to facilitate the labors of their bees, you might suppose that they imagined these delicate insects to be possessed of nerves of steel and sinews of iron or adamant; or else that they took them for miniature locomotives, always fired up and capable of an indefinite amount of exertion. A bee *cannot* put forth more than a certain amount of physical exertion, and if a large portion of this is spent in absolutely fighting against difficulties, from which it might easily be guarded, it must be very obvious to any one who thinks on the subject at all, that a great loss must be sustained by its owner.

If some of these thoughtless owners returning home with a heavy burden, were compelled to fall down stairs half a dozen times before they could get into the house, they might perhaps think it best to guard their industrious workers against such discouraging accidents. If bees are tossed violently about by the winds, as they attempt to enter their hives, they are often fatally injured, and the whole colony so *discouraged*, to say nothing more, that they do not gather near so much as they otherwise would.

The arrangement of my Protector is such that the bees, if blown down, fall upon a sloping bank of soft grass, and are able to enter the hives without much inconvenience.

Just as soon as our cultivators can be convinced, by practical results, that bee-keeping, for the capital invested, may be made a most profitable branch of rural economy, they will see the importance of putting their bees into suitable hives, and of doing all that they can, to give them a fair chance; until then, the mass of them will follow the beaten track, and attribute their ill success, not to their own ignorance, carelessness or stupidity, but to their want of "luck," or to the overstocking of the country with bees. I hope, before many years, to see the price of good honey so reduced that the poor man can place it on his table and feast upon it, as one of the cheapest luxuries within his reach.

On page 20, a statement was given of Dzierzon's experience as to the profits of bee-keeping. The section of country in which he resides, is regarded by him as unfavorable to Apiarian pursuits. I shall now give what I consider a safe estimate for almost any section in our country; while in unusually favorable locations it will fall far below the results which may be attained. It is based upon the supposition that the bees are kept in properly constructed hives so as to be strong early in the season, and that the increase of stocks is limited to one new one from two old ones. Under proper management, one year with another, about ten dollars worth of honey may be obtained for every two stocks wintered over. The worth of the new colonies, I set off as an equivalent for labor of superintendence, and interest on the money invested in bees, hives, fixtures, &c.

A careful, prudent man who will enter into bee-keeping moderately at first, and extend his operations only as his skill and experience increase, will, by the use of my hives, find that the preceding estimate is not too large. Even on the ordinary mode of bee-keeping, there are many who will consider it rather below than above the mark. If thoroughly careless persons are determined to "try their luck," as they call it, with bees, I advise them by all means, in mercy to the bees, to adopt the non-swarming plan. Improved methods of management with such persons will be of little or no use, unless you could improve their habits first, and very often their brains too! Every dollar that such persons spend upon bees, unless with the slightest possible departure from the old-fashioned plans, is a dollar worse than thrown away. In those parts of Europe where bee-keeping is carried on upon the largest scale, the mass adhere to the old system; this they understand, and by this they secure a certainty, whereas in our country, thousands have been induced to enter upon the wildest schemes, or at least to use hives which could not furnish them the very information needed for their successful management. A simple box furnished with my frames, will enable the masses, without departing materially from the common system, to increase largely the yield from their bees.

In addition to the information given in the Introduction, respecting the success of Dzierzon's system of management, I have recently ascertained that one of its ablest opponents in Germany, has become thoroughly convinced of its superior value. The Government of Norway has appropriated $300, per annum, for the ensuing three years, towards diffusing a knowledge of Dzierzon's method, in that country; having previously despatched Mr. Hanser, Collector of Customs, to Silesia to visit Mr. Dzierzon, and acquire a practical knowledge of his system of management. He is now employed in distributing model hives, in the provinces, and imparting information on improved bee-culture.

Note.—The time has hardly come when the attention of any of our State authorities can be attracted to the importance of bee-culture. It is only of late that they have seemed to manifest any peculiar interest in promoting the advancement of agricultural pursuits. A Department of Agriculture ought to have been established, years ago, by the National Government at Washington. Let us hope that the Administration now in power, will establish a lasting claim to the gratitude of posterity, by taking wise and efficient steps to advance the agricultural interests of the country. A National Society to promote these interests has recently been established, and much may be hoped from its wisdom and energy. Until some disinterested tribunal can be established, before which all inventions and discoveries can be fairly tested, honest men will suffer, and ignorance and imposture will continue to flourish. Lying advertisements and the plausible misrepresentations of brazen-faced impostors, will still drain the purses of the credulous, while thousands, disgusted with the horde of impositions which are palmed off upon the community, will settle down into a dogged determination to try nothing new. A society before which every thing, claiming to be an improvement in rural economy, could be fairly tested, would undoubtedly be shunned by ignorant and unprincipled men, who now find it an easy task to procure any number of certificates, but who dread nothing so much as honest and intelligent investigation. The reports of such a society after the most thorough trials and examinations, would inspire confidence, save the community from severe losses, and encourage the ablest minds to devote their best energies to the improvement of agricultural implements.

CHAPTER XVIII. THE ANGER OF BEES. REMEDY FOR THEIR STING. BEE-DRESS. INSTINCTS OF BEES.

If the bee was disposed to use, without any provocation, the effective weapon with which it has been provided, its domestication would be entirely out of the question. The same remark however, is equally true of the ox, the horse or the dog. If these faithful servants of man were respectively determined to use, to the very utmost their horns, their heels and their teeth, to his injury, he would never have been able to subject them to his peaceful authority. The gentleness of the honey-bee, when kindly treated, and managed by those who properly understand its instincts, has in this treatise been frequently spoken of, and is truly astonishing. They will, especially in swarming time, or whenever they are gorged with honey, allow any amount of handling which does not hurt them, without the slightest show of anger. For the gratification of others, I have frequently taken them up, by handfuls, suffered them to run over my face, and even smoothed down their glossy backs as they rested on my person! Standing before the hives, I have, by a rapid sweep of my hands, caught numbers of them at once, just as though they were so many harmless flies, and allowed them, one by one, to crawl out, by the smallest opening, to the light of day; and I have even gone so far as to imitate many of the feats which the celebrated English Apiarian, Wildman, was accustomed to perform; who having once secured the queen of a hive, could make the bees cluster on his head, or hang, like a flowing beard, in large festoons, from his chin. Wildman, for a long time, made as great a mystery of his wonderful performances, as the spirit-rappers of the present day, do of theirs; but at last, he was induced to explain his whole mode of procedure; and the magic control which he possessed over the bees, and which was, by the ignorant, ascribed to his having bewitched them, was found to be owing entirely to his superior acquaintance with their instincts, and his uncommon dexterity and boldness.

"Such was the spell, which round a Wildman's arm Twin'd in dark wreaths the fascinated swarm; Bright o'er his breast the glittering legions led, Or with a living garland bound his head. His dextrous hand, with firm yet hurtless hold, Could seize the chief, known by her scales of gold,

Prune 'mid the wondering train her filmy wing, Or o'er her folds the silken fetter fling." *Evans.*

M. Lombard, a skillful French Apiarian narrates the following interesting occurrence, which shows how peaceable bees are in swarming time, and how easily managed by those who have both skill and confidence.

"A young girl of my acquaintance," he says, "was greatly afraid of bees, but was completely cured of her fear by the following incident. A swarm having come off, I observed the queen alight by herself at a little distance from the Apiary. I immediately called my little friend that I might show her the queen; she wished to see her more nearly, so after having caused her to put on her gloves, I gave the queen into her hand. We were in an instant surrounded by the whole bees of the swarm. In this emergency I encouraged the girl to be steady, bidding her be silent and fear nothing, and remaining myself close by her; I then made her stretch out her right hand, which held the queen, and covered her head and shoulders with a very thin handkerchief. The swarm soon fixed on her hand and hung from it, as from the branch of a tree. The little girl was delighted above measure at the novel sight, and so entirely freed from all fear, that she bade me uncover her face. The spectators were charmed with the interesting spectacle. At length I brought a hive, and shaking the swarm from the child's hand, it was lodged in safety, and without inflicting a single wound."

The indisposition of bees to sting, when swarming, is a fact familiar to every practical bee-keeper: but I have not in all my reading or acquaintance with Apiarians, ever met with a single observation which has convinced me that the philosophy of this strange fact was thoroughly understood. As far as I know, I am the only person who has ever ascertained that when bees are filled with honey, they lose all disposition to volunteer an assault, and who has made this curious law the foundation of an extensive and valuable system of practical management. It was only after I had thoroughly tested its universality and importance, that I began to feel the desirableness of obtaining a perfect control over each comb in the hive; for it was only then that I saw that such control might be made available, in the hands of any one who could manage bees in the ordinary way. The result of my whole system, is to make the bees unusually gentle, so that they are not only peaceable when any necessary operation is being performed, but at all other times. Even if I could open hives and safely manage at pleasure, still if the result of such proceedings was to leave the bees in an excited state, so as to make them unusually irritable, it would all avail but very little.

There is, however, one difficulty in managing bees so as not to incur the risk of being stung at all, which attaches to every system of bee-culture. If an Apiary is approached when the bees are out in great numbers, thousands and tens of thousands will continue their busy pursuits without at all interfering with those who do not molest them. Frequently, however, there will be a few cross bees which come buzzing around our ears, and seem determined to sting without the very slightest provocation. From such lawless bees no person without a bee-dress is absolutely safe. By repeated examinations I have ascertained that *disease* is the cause of such unreasonable irritability. I am never afraid that a healthy bee will attack me unless unusually provoked; and am always sure as soon as I hear one singing about my ears that it is incurably diseased. If such a bee is dissected it will be found to exhibit the unmistakable evidence that a peculiar kind of dysentery has already fastened upon its system. In the first stages of this complaint the insect is very irritable, refuses to labor, and seems unable or unwilling to distinguish friend from foe. As the disease progresses, it becomes stupid, its body swells up, and is filled with a great mass of yellow matter, and being unable to fly, it crawls on the ground, in front of the hive, and speedily perishes. I have never been able to ascertain the cause of this singular malady, nor can I suggest any remedy for it. I hope that some scientific Apiarians will investigate it closely, for if it could only be remedied, we might have hundreds of colonies on our premises and in our gardens, and yet be perfectly safe.

A person thoroughly acquainted with the leading principles of bee-culture as they are set forth in this Manual, will *never under any circumstances* find it necessary to provoke to fury a colony of bees. Let it be remembered that nothing can be more terribly vindictive than a family of bees when thoroughly aroused by gross abuse or unskillful treatment. Let their hive be suddenly overthrown or violently jarred, or let them be provoked by the presence of a sweaty horse, or any animal offensive to them, so that the anger at first manifested by a few, is extended to the whole community, and the most severe and sometimes dangerous consequences may ensue. In the same way in the management of the animals most useful to man, by ignorance or abuse, they may be roused to a state of frantic desperation, and limbs may be broken, and often lives destroyed;

and yet no one possessed of common sense, attributes such calamities, except in very rare instances, to any thing else than carelessness or want of skill. Let it be remembered that even the most peaceable stock of bees can, in a very few days, by abusive treatment be taught to look on every living thing as an enemy, and to sally forth with the most spiteful intentions, as soon as any one approaches their domicile. How often does it happen that the vicious beast, which its owner so passionately belabors, is far less to blame for its obstinacy, than the equally vicious brute who so unmercifully beats it!

A word now to those timid females who are almost ready to faint, or to go into hysterics if a bee enters the house, or approaches them in the garden or fields. Such alarm is entirely uncalled for. It is only in the vicinity of their homes, and in resistance to what they consider an evil design upon their very altars and firesides that these insects ever volunteer an attack. Away from home, they are as peaceably inclined as you could desire. If you attack them, they are much more eager to escape than to offer you any annoyance, and they can be induced to sting, only when they are compressed, either by accident or design.

Let not any of my readers think that they have even a slight encouragement, from this conduct of the bee, to reserve all their sweet smiles and honied words for the world abroad, while they give free vent, in the sacred precincts of home, to ill-natured looks and ill-tempered language; for towards the occupants of its honied dome, the bee is all kindness and affection. In the experience of many years I never saw an instance in which two bees, members of the same family, ever seemed to be actuated by any but the very kindest feelings toward each other. In their busy haste they often jostle against each other, but where every thing is well meant, every thing is well received: tens of thousands all live together in the sweetest harmony and peace, when very often if there are only two or three children in a family, the whole household is tormented by their constant bickerings and contention. Among the bees the good mother is the honored queen of her happy family; they all wait upon her steps with unbounded reverence and affection, make way for her as she moves over the combs, smooth and brush her beautiful plumes, offer her food from time to time, and in short do all that they possibly can to make her perfectly happy; while too often children treat their mothers with irreverence or neglect, and instead of striving with loving zeal to lighten their labors and save their steps, they treat them more as though they were servants hired only to wait upon every whim and to humor every caprice.

Let us pause for a moment, and contemplate further the admirable arrangement by which the instinct of the bee which disposes it to defend its treasures, is made so perfectly compatible with the safety both of man and the domestic animals under his care. Suppose that away from home, bees were as easily provoked, as they are in the immediate vicinity of their hives, what would become of our domestic animals among the clover fields in the pastures? A tithe of the merry gambols they now so safely indulge in, would speedily bring about them a swarm of these infuriated insects. In all our rambles among the green fields, we should constantly be in peril; and no jocund mower would ever whet his glittering scythe, or swing his peaceful weapon, unless first clad in a dress impervious to their stings. In short, the bee, instead of being the friend of man, would be one of his most vexatious enemies, and as has been the case with the wolves and the bears, every effort would be made for their utter extermination.

The sting of a bee often produces very painful, and upon some persons, very dangerous effects. I am persuaded, from the result of my own observation, that the bee seldom stings those whose systems are not sensitive to its venom, while it seems to take a special and malicious pleasure in attacking those upon whom it produces the most painful effects! It may be that something in the secretions of such persons both provokes the attack, and causes its consequences to be more severe.

I should not advise persons upon whose system the sting of a bee produces the most agonizing pain, and violent, if not dangerous symptoms, to devote any attention to the practical part of an Apiary; although I am acquainted with a lady who is thus severely affected, and who yet, strange to say, is a great enthusiast in Apiarian pursuits! I have met with individuals, upon whom a sting produced the singular effect of causing their breath to smell like the venom of the enraged insect! The smell of the poison resembles almost perfectly that of a ripe banana. It produces a very irritating effect upon the bees themselves; for if a minute drop of it is extended to them, on a stick, they at once manifest the most decided anger.

It is well known that the bee is a lover of sweet odors, and that unpleasant ones are very apt to excite its anger. And here I may as well speak plainly, and say that bees have a special dislike to persons whose habits are not cleanly, and particularly to those who bear about them, a perfume not in the very least resembling those

"Sabean odors From the spicy shores of Araby the blest,"

of which the poet so beautifully discourses. Those who belong to the family of the "great unwashed," will find to their cost that bees are decided foes to all of their tribe. The peculiar odor of some persons, however cleanly, may account for the fact that the bees have such a decided antipathy to their presence, in the vicinity of their hives. It is related of an enthusiastic Apiarian, that after a long and severe attack of fever, he was never able to take any more pleasure in his bees; his secretions seem to have undergone some change, so that the bees assailed him as soon as he ventured to approach their hives.

Nothing is more offensive to bees than the impure breath exhaled from human lungs; it excites them at once to fury. Would that in their hatred for impure air, human beings had only a tithe of the sagacity exercised by bees! It would not be long before the thought of breathing air loaded with all manner of impurities from human lungs, to say nothing of its loss of oxygen, would excite unutterable loathing and disgust.

As the smell of a sweaty horse is very offensive to the bees, it is never safe to allow these animals to go near a hive, as they are sometimes attacked and killed by the furious insects. Those engaged in bee-culture on a large scale, will do well to enclose their Apiaries with a strong fence, so as to prevent cattle from molesting the hives. If the Apiary is enclosed by a high fence, with sharp and strong pickets, and the door is furnished with a strong lock, it will prevent the losses which in some localities are so common from human pilferers. Such losses may be guarded against, by fastening a wrought iron ring into the top of each hive, well clinched on the inside; an iron rod may run through these rings, and thus with two padlocks and fixtures, (one at each end,) a dozen or more hives may be secured. I am happy to say that in most localities such precautions are entirely unnecessary. A place in which the stealing of honey and fruit is practiced by any except those who are candidates for State's Prison, is in a fair way of being soon considered as a very undesirable place of residence. If owners of Apiaries, gardens and orchards, could be induced to pursue a more liberal policy, and not be so meanly penurious as they often are, I am persuaded that they would find it conduce very highly to their interests. The honey and fruit expended with a cheerful, hearty liberality, would be more than repaid to them in the good will secured, and in the end would cost much less than bars and bolts. Reader! do not imagine that I have the least idea that a thoroughly selfish man, can ever be made to practice this or any other doctrine of benevolence. Demonstrate it again and again, until even to his narrow and contracted view it seems almost as clear as light, still he will never find the heart to reduce it to practice. You might almost as well expect to transform an incarnate fiend into an angel of light, by demonstrating that "Wisdom's ways are ways of pleasantness," while "the path of the transgressor is hard," as to attempt to stamp upon a heart encrusted with the adamant of selfishness, the noble impress of a liberal spirit.

Of all the senses, that of smell in the bee, seems to be the most perfect. Huber has demonstrated its exceeding acuteness, by numerous interesting experiments. If honey is placed in vessels from which the odor can escape, but in which it cannot be seen, the bees will soon alight upon them and eagerly attempt to find an entrance. It is by this sense, unquestionably, that they recognize the members of their own community, although it seems to us very singular that each colony should have its own peculiar scent. Not only can two colonies be safely united by giving them the same odor, but in the same way any number of colonies may be made to live in perfect peace. If hundreds of hives are all connected by gauze wire ventilators, so that the air passes freely from one to another, the bees will all live in absolute harmony, and if any bee attempts to enter the wrong hive, he will not be molested. The same result can often be attained by feeding colonies from a common vessel. I have seen literally hundreds of thousands of bees that after being treated in this way so as to acquire the same odor, were always gentle towards each other, while if a single bee from a strange Apiary, lit upon the feeder, it was sure to be killed.

I have described, (p. 213,) the use which I make of peppermint, in order to prevent bees from quarreling when they are united. The Rev. Mr. Kleine, in a recent number of the Bienenzeitung, has recommended the use of another article, which he finds to be very useful in preventing robbing. His statement would have come in more appropriately in the Chapter on Robbing, but was not received until too late. He says that the most convenient and effectual mode of arresting and repelling the attacks of robbers, is, to impart to the attacked hive some intensely powerful and unaccustomed odor. He effects this most readily, by placing a small portion of *musk* in the attacked hive, late in the evening, when all the robbers have retreated. On the following morning, the bees, (provided they have a healthy queen,) will promptly and boldly meet their assailants, and these in turn are non-plussed by the unwonted odor, and if any of them enter the hive and carry off some of the coveted booty, they will not be recognized nor received at home on their return, on account of their strange smell, but will be at once seized as strangers, and killed by their own household. Thus the robbing is speedily brought to a close.

In combination with my blocks, this device might be made very effectual. When the Apiarian perceives that a hive is being robbed, let him shut up the entrance: before dusk he can open it and allow the robbers to go home, and then: put in a small piece of musk: the entrance next day may be kept so contracted that only a single bee can enter at once. In the union of stocks the same substance might be used advantageously. A short time before the process is attempted, each colony might have a small dose of musk (a piece of musk tied up in a little bag,) and they would then be sure to agree. I prefer, however, in most cases, the use of scented sugar-water.

By using my double hives, and putting a small piece of gauze-wire on an opening made in the partition, the two colonies having the same scent will always agree; this will be very convenient where they are compelled to live as such near neighbors, and enables the Apiarian at any time to unite them and appropriate their surplus stores. These double hives are admirably adapted to the wants of those who wish to make the smallest possible departure from the old system, as they need make no change, except to unite the stocks instead of killing the bees.

I have already remarked that no operation should ever be attempted upon bees, by which a whole colony is liable to be excited to an ungovernable pitch of fury. Such operations are *never* necessary; and a skillful Apiarian will, by availing himself of the principles laid down in this Treatise, both easily and safely do everything that is at all desirable, even to the driving of a powerful colony from an old box hive. When bees are improperly dealt with, they will "compass" their assailant "about," with the most savage ferocity, and woe be to him if they can creep up his clothes, or find on his person a single unprotected spot! On the contrary, when not provoked by foolish management or wanton abuse, the few who are bent on mischief, appear to retain still some touch of grace, amid all their desperation. Like the thorough bred scold, who by the elevated pitch of her voice, often gives timely warning to those who would escape from the sharp sword of her tongue, a bee bent upon mischief raises its note almost an octave above the peaceable pitch, and usually gives us timely warning, that it means to sting, if it can. Even then, it will seldom proceed to extremities, unless it can leave its sting somewhere upon the face of its victim, and usually as near as possible to the eye; for bees and all other members of the stinging tribe, seem to have, as it were, an intuitive perception that this is the most vulnerable spot upon the "human face divine." If the head is quietly lowered, and the face covered with the hands, they will often follow a person for some rods, all the time sounding their war note in his ears, taunting him for his sneaking conduct, and daring him, just for one single moment, to look up and allow them to catch but a glimpse of his coward face!

If a person is suddenly attacked by angry bees, no matter how numerous or vindictive they may be, not the slightest attempt should ever be made to act on the offensive. If a single bee is violently struck at, a dozen will soon be on hand to avenge the insult, and if the resistance is still continued, hundreds and at last thousands will join in the attack. The assailed party should quickly retreat from the vicinity of the hives, to the protection of a building, or if none is near, he should hide himself in a clump of bushes, and lie perfectly still, with his head covered, until the bees leave him.

REMEDIES FOR THE STING OF A BEE.

If only a few of the host of remedies, so zealously advocated, could be made effectual, few persons would have much reason to dread being stung. Most of them, however, are of no manner of use whatever. Like the prescriptions of the quack, they are absolutely worse than doing nothing at all.

The first thing to be done after being stung, is to pull the sting out of the wound *as quickly as possible*. Even after it is torn from the body of the bee, the muscles which control it, are in active operation, and it penetrates deeper and deeper into the flesh, injecting continually more and more of its poison into the wound. Every Apiarian should have about his person, or close at hand, a small piece of looking-glass, so that he may be able with the least possible delay to find and remove a sting. In most cases if it is at once removed, it will produce no serious consequences; whereas if suffered to empty all its vials of wrath, it may cause great inflammation and severe suffering. After the sting is removed, the utmost possible care should be taken, not to irritate the wound by the very *slightest rubbing*. However intense the smarting, and of course the disposition to apply friction to the wound, it should never be done, as the poison will at once be carried through the circulating system, and severe consequences may ensue. As most of the popular remedies are rubbed in, they are of course worse than nothing. Be careful not to *suck* the wound as so many persons do; this produces irritation in the same way with rubbing. Who does not know that a musquito bite, even after the lapse of several days, may be brought to life again, by violent rubbing or sucking? The moment that the blood is put into a violent and unnatural circulation, the poison is quickly diffused over a considerable part of the system. If the mouth is applied to the wound, other unpleasant consequences may ensue. While the poison of most snakes and many other noxious animals affects only the circulating system, and may therefore be swallowed with impunity, the poison of the bee acts powerfully, not only upon the circulating system, but upon the organs of digestion. The most distressing head-aches are often produced by it.

From my own experience, I recommend *cold water* as the very best remedy with which I am acquainted, for the sting of a bee. It is often applied in the shape of a plaster of mud, but may be better used by wetting cloths and holding them gently to the wound. Cold water seems to act in two ways. The poison of the bee being very volatile, is quickly dissolved in water; and the coldness of the water has also a powerful tendency to check inflammation and to prevent the virus from being taken up by the absorbents and carried through the system. The leaves of the plantain, crushed and applied to the wound, will answer as a very good substitute when water cannot at once be procured. The broad-leafed plantain, or as some call it, "the toad plantain," is regarded by many as possessing a very great efficacy. Bevan recommends the use of spirits of hartshorn, applied to the wound, and says that in cases of severe stinging its internal use is beneficial. Whatever remedy is applied, should be used if possible, without a moment's delay. The immediate extraction of the sting, will be found, even if nothing more is done, much more efficacious than any remedy that can be applied, after it has been allowed to remain and discharge all its venom into the wound.

It may be some comfort to those who are anxious to cultivate bees, to know that after a while the poison will produce less and less effect upon their system. When I first became interested in bees, a sting was quite a formidable thing, the pain often being very intense, and the wound swelling so as sometimes to obstruct my sight. At present, the pain is usually slight, and if I can only succeed in quickly extracting the sting, no unpleasant consequences ensue, even if no remedies are used. Huish speaks of seeing the bald head of Bonner, a celebrated practical Apiarian, lined with bee stings which seemed to produce upon him no unpleasant effects. Like Mithridates, king of Pontus, he seemed almost to thrive upon poison itself!

I have met with a highly amusing remedy very gravely propounded by an old English Apiarian. I mention it more as a matter of curiosity, than because I imagine that any of my readers will be likely to make trial of it. He says, let the person who has been stung, catch as speedily as possible, another bee, and make it sting on the same spot! It requires some courage even in an enthusiastic disciple of Huber, to venture upon such a singular homeopathic remedy; but as this old writer had previously stated that the oftener a person was stung, the less he suffered from the venom, and as I had proved, in my own experience, the truth of this assertion, I determined to make trial of his remedy. I allowed a bee to sting me upon the finger and suffered the sting to remain until it had discharged all its venom. I then compelled another bee to insert its sting as near as possible in the same spot. I used no remedies of any kind, and had the satisfaction, in my zeal for new discoveries, of suffering more from the pain and swelling, than I had previously experienced for years.

An old writer recommends a powder of dried bees, for distressing cases of stoppages; and some of the highest medical authorities have recently recommended a tea made by pouring boiling water upon bees, for the same complaint, while the homeopathic physicians employ the poison of the bee, which they call *apis*, for a great variety of maladies. That it is capable of producing intense head-aches any one who has been stung, or who has tasted the poison, very well knows.

BEE-DRESS.

Timid Apiarians, and all who are liable to suffer severely from the sting of a bee, should by all means furnish themselves with the protection of a bee-dress. The great objection to gauze-wire veils or other materials of which such a dress has been usually made, is that they obstruct clear vision, so highly important in all operations, besides producing such excessive heat and perspiration, as to make the Apiarian peculiarly offensive to the bees. I prefer to use what I shall call a *bee-hat*, of entirely novel construction. It is made of wire cloth, the meshes of which are too fine to admit a bee, and yet coarse enough to allow a free circulation of air, and to permit distinct sight. The wire cloth should first be fastened together in a circular shape, like a hat, and large enough to go very easily over the head; its top may be of cotton cloth, and it should have the same material fastened around its lower edge, and furnished with strings to draw it so closely around the neck and shoulders that a bee cannot creep under it. Woolen stockings may then be drawn over the hands, or better still, India Rubber gloves, such as are now in very common use, may be worn; these gloves are impenetrable to the sting of a bee, and yet are so soft and pliant as scarcely in the least to interfere with the operations of the Apiarian.

If it were not for the diseased bees of which I have several times spoken, such precautions would be entirely unnecessary. The best Apiarians as it is, dispense with them, even at the cost of a sting now and then

INSTINCTS OF BEES.

This treatise has already grown to such a length, that I must be exceedingly brief on a point peculiarly interesting to all who delight in investigating the wonders of the insect world. In the preceding parts of the work, numerous proofs have been given of the refined instincts of the bee. It is impossible always to draw the line between instinct and reason, and very often some of the actions of animals and insects appear to be the result of a process of reasoning apparently almost the same with the exercise of the reasoning faculty in man. "There is this difference" says Mr. Spence, "between intellect in man, and the rest of the animal creation. Their intellect teaches them to follow the lead of their senses, and to make such use of the external world as their appetites or instincts incline them to,—and *this is their wisdom*: while the intellect of man, being associated with an immortal principle, and connected with a world above that which his senses reveal to him, can, by aid derived from Heaven, control those senses, and render them obedient to the governing power of his nature; and *this is his wisdom.*"

This subject has seldom been more happily expressed than by Mr. Spence. The line of distinction between man and the lower orders of creation, is not the mere fact that he reasons and they do not, but that he has a moral and accountable nature, while they have nothing of the kind.

"It will be evident," says Bevan, "that though I make a distinction between the instinct and the reason of bees, I do not confound their reason with the reason of man. But to obviate all possibility of misconception, I will at once define my meaning, when I use the terms insect reason and instinct."

"By *reason*, I mean the power of making deductions from previous experience or observation, and thereby of adapting means to ends. *Instinct* I regard as a disposition and power to perform certain actions in the same uniform manner, depending upon nice mechanism and having no reference either to observation or experience; operating on the means, without anticipation of the end, incited by no hope, controlled by no foreboding. Those who have attended to this subject, will be aware that *insect reason*, as above defined, is more restricted in its functions than *the reason of man*; to which is superadded the power of distinguishing between the true and the false, and, according to some metaphysicians, between right and wrong. Reason, in man, has a regular growth and a slow progression; all the arts he practices evince skill and dexterity, proportioned to the pains which have been taken in acquiring them. In the lower links of creation, but little of this gradual improvement is observable; their powers carry them almost directly to their object. They are perfect, as Bacon says, in all their members and organs from the very beginning."

"Far different Man, to higher fates assign'd, Unfolds with tardier step his Proteus mind, With numerous Instincts fraught, that lose their force Like shallow streams, divided in their course; Long weak, and helpless, on the fostering breast, In fond dependence leans the infant guest, Till reason ripens what young impulse taught, And builds, on sense, the lofty pile of thought; From earth, sea, air, the quick perceptions rise, And swell the mental fabric to the skies." *Evans.*

I shall here narrate a very remarkable instance of sagacity which seems to approach as near to human reason, as any thing in the bee which has ever fallen under my notice. In the year 1851, I had a small model hive constructed, into which I temporarily placed a swarm of bees. The particular object which I had in view, was to test the feasibility of some plans which I had recently devised, for facilitating the storing of honey in small tumblers. The bees, in a short time, filled the hive and stored about a dozen glasses with honey. I was called away from them, for a few days, and was much surprised, on my return, to find that the honey which had been stored up in the hive and sealed over for Winter use, was all gone, and the cells filled with eggs and young worms! The hive stood in a covered bee house, and the bees had built a large quantity of comb on the *outside* of the hive, into which they had transferred the honey taken from the interior. The object of this unusual procedure was, beyond all question, to give the poor queen a place within the hive for laying her eggs: for this purpose they uncapped and emptied all the cells so carefully sealed over, instead of using the new comb on the outside for the brood.

Those who wish to study the Natural History of the honey-bee, to the best advantage, will derive great aid in their investigations, from the use of my *Observing Hives*. Each comb in these hives is attached to a movable frame, and they all admit of easy removal. In this respect the construction of the hive is entirely new, and while it greatly facilitates the business of observation, it enables the Apiarian, on the approach of cool weather, to transfer his bees from a hive in which they cannot winter, to one of the common construction. As soon as the weather in the Spring is sufficiently warm, they may again be placed in the observing hive, in which, (as both sides of every comb admit of inspection,) every bee can be seen, and all the wonders of the hive are exposed to the full light of day; In the common observing hives experiments are often conducted with great difficulty, by cutting away parts of the comb, whereas in mine, they can all be performed by the simple removal of one of the frames, and if the colony becomes reduced in numbers, it may, in a few moments, be strengthened by helping it to maturing brood from one of the other hives. A very intelligent writer in a description of the different hives exhibited at the World's Fair, in London, lamented that no method had yet been devised of enabling bees to cluster, in cool weather, in an observing hive, and that it was found next to impossible to preserve them in such hives over Winter. By the use of the movable frames, this difficulty is entirely obviated.

I cannot allow this work to come to a close, without acknowledging my great obligations to Mr. Samuel Wagner, of York, Pennsylvania. To him I am indebted for a knowledge of Dzierzon's discoveries, and for many valuable suggestions scattered throughout the Treatise.

FOOTNOTES:

[1] The author of this work regrets that his experience does not enable him to speak with such absolute confidence as to the character of all the bee keepers whom he has known.

[2] In this way she is sure to deposit the egg in the cell she has selected.

[3] If ever there lived a genuine naturalist, Swammerdam was the man. In his History of Insects, published in 1737, he has given a most beautiful drawing of the ovaries of the queen bee. The sac which he supposed secreted a fluid for sticking the eggs to the base of the cells is the seminal reservoir or spermatheca.

[4] Bevan.

[5] This work being intended chiefly for practical purposes, I have thought best to use, as little as possible, the technical terms and minute anatomical descriptions of the scientific entomologist.

[6] Bevan.

[7] Having already spoken of Swammerdam, I shall give a brief extract from the celebrated Dr. Boerhaave's memoir of this wonderful naturalist, which should put to the blush, if any thing can, the arrogance of those superficial observers who are too wise in their own conceit, to avail themselves of the knowledge of others.

"This treatise on Bees proved so fatiguing a performance, that Swammerdam never afterwards recovered even the appearance of his former health and vigor. He was almost continually engaged by day in making observations, and as constantly engaged by night in recording them by drawings and suitable explanations."

"This being summer work, his daily labor began at six in the morning, when the sun afforded him light enough to survey such minute objects; and from that hour till twelve, he continued without interruption, all the while exposed in the open air to the scorching heat of the sun, bareheaded for fear of intercepting his sight, and his head in a manner dissolving into sweat under the irresistible ardors of that powerful luminary. And if he desisted at noon, it was only because the strength of his eyes was too much weakened, by the extraordinary afflux of light and the use of microscopes, to continue any longer upon such small objects, though as discernible in the afternoon, as they had been in the forenoon."

"Our author, the better to accomplish his vast, unlimited views, often wished for a year of perpetual heat and light to perfect his inquiries, with a polar night to reap all the advantages of them by proper drawings and descriptions."

[8] The formation of swarms will be particularly described in another chapter.

[9] Suppose that we are unable to give a satisfactory answer to any of these questions, does our ignorance on these points disprove the *fact* of the existence of such a jelly?

[10] Bevan.

[11] Some very extraordinary instances are related of the protraction of life in snails. After they had lain in a cabinet above fifteen years, immersing them in water caused them to revive and crawl out of their shells.

[12] A writer in the New England Farmer for March, 1853, estimates that the mild winter has been worth in the saving of fodder to the farmers of New Hampshire alone, two and a half millions of dollars! By suitable arrangements, bees even in the very coldest climates can have all the advantages of a mild winter.

[13] The cost of the glass for one hive so as to give the air space all around, if purchased at the wholesale prices will not exceed 25 cts. Where three hives are made in one structure, the glass for the three will cost less than 50 cents; if double glass is

not used, the expense would be less by one half.

[14] The observations to test the temperature of the Protector were made in Greenfield, Massachusetts, in latitude 42 deg. 36 min.

[15] The beautiful open or Franklin stoves, manufactured by Messrs. Jagger, Treadwell and Perry, of Albany, deserve the highest commendation: they economize fuel as well as life and health.

[16] Dr. Scudamore, an English physician who has written a small tract on the formation of artificial swarms, says that he once knew "as many as ten swarms go forth at once, and settle and mingle together, forming literally a monster meeting!" Instances are on record of a much larger number of swarms clustering together. A venerable clergyman, in Western Massachusetts, related to me the following remarkable occurrence. In the Apiary of one of his parishioners, five swarms lit in one mass. As there was no hive which would hold them, a very large box was roughly nailed together, and the bees were hived in it. They were taken up by sulphur in the Fall, when it was perfectly evident that the five swarms had occupied the same box as independent colonies. Four of them had commenced their works, each one near a corner, and the fifth one in the middle, and there was a distinct interval separating the works of the different colonies. In Cotton's "My Bee Book," there is a cut illustrating a hive in which two colonies had built in the same manner.

[17] I have often spent more than ten minutes in opening and shutting a single frame in the Huber hive, and even then, have sometimes crushed some of the bees.

[18] The scent of the hives, during the height of the gathering season, will usually inform us from what sources the bees have gathered their supplies.

[19] If they cannot obtain it, the Apiarian must himself furnish it.

[20] The queens taken from such hives may be advantageously used in forming artificial colonies.

[21] Bevan.

[22] Bevan.

[23] A bee, a few days after it is hatched, is as fully competent for all its duties, as it ever will be, at any subsequent period of its life.

[24] Report on bees to the Essex County Agricultural Society, 1851.

[25] Instead of using sticks, I much prefer to make the drumming with the open palms of my hands.

[26] The bees in each colony had probably contracted the same smell, and could not distinguish friends from foes.

Index

www.ingramcontent.com/pod-product-compliance
Lightning Source LLC
Chambersburg PA
CBHW060755100426
42813CB00004B/828